SO-AIZ-262

American Railroads

John F. Stover

American Railroads

SECOND EDITION

THE UNIVERSITY OF CHICAGO PRESS
CHICAGO & LONDON

JOHN F. STOVER is Professor Emeritus of History at Purdue University and
the author of several books, including *The Life and Decline of the American Railroads* (1971); *History of the Illinois Central Railroad* (1975); *Iron Road to the West*
(1978); and *History of the Baltimore and Ohio Railroad* (1987).

The University of Chicago Press, Chicago 60637
The University of Chicago Press, Ltd., London
© 1961, 1997 by The University of Chicago
All rights reserved.
Second Edition published 1997
Printed in the United States of America
06 05 04 03 02 01 00 99 98   2 3 4 5
Title page illustration courtesy of the Albany Institute of History and Art

Library of Congress Cataloging-in-Publication Data

Stover, John F.
    American railroads / John F. Stover. — 2nd ed.
        p.   cm. — (The Chicago history of American civilization)
    Includes bibliographical references and index.
    ISBN 0-226-77657-3 (cloth : alk. paper). — ISBN 0-226-77658-1 (pbk. : alk.
        paper)
    1. Railroads—United States—History.   I. Series.
HE2751.S7   1997
385'.0973—dc21   97-449
                                                                          CIP

⊗ The paper used in this publication meets the minimum requirements of the
American National Standard for Information Sciences—Permanence of Paper for Printed Library Materials, ANSI Z39.48-1984.

For Charry, Bob, and John

Contents

Illustrations

Illustrations

Editor's Foreword to the Second Edition

As Mr. Stover explains in this book, since their appearance on the American scene railroads have played a decisive role in nearly every major movement in American history. And he succinctly explains how those roles have changed with our technology and our economy. In this revised and updated edition, he has displaced some chapters in the earlier edition, added a new map and new data, and updated and expanded the Suggested Reading. His new chapters survey the new age of railroads, the new problems and opportunities, mergers, and legislation. The need for this edition itself illustrates the dynamism of American history, and shows us how progress produces the challenges of obsolescence. If the recent history of American railroads lacks the expansive romantic drama of the earlier history, it does remain a touchstone of the flexibility of American institutions. Mr. Stover reminds us of the recurring problem in a nation of ever-advancing technology, of making the most of the enormous investment in earlier technologies.

When an eminent British railroad builder, Edward Watkin, visited the United States in the mid-nineteenth century, he was appalled at how hastily and flimsily the Americans laid their tracks and built their bridges. He contrasted them with the sturdy British lines which he said were built "to last for ever." Americans, he complained, somehow did not realize that railroads were "a final improvement in the means of locomotion." As Mr. Stover continues his account of the later history of American transportation in this new edition, he alerts our imaginations to the unpredictability of innovation, and the need to keep our institutions adaptable to the unexpected.

DANIEL J. BOORSTIN

Editor's Foreword to the First Edition

There are few more distinctively American subjects than the history of American railroads. While other nations too have had a railroad history, our nation of vast continental spaces has been decisively shaped by its means of transportation. Future historians may well divide American life since the early nineteenth century into an Age of the Railroad, followed by an Age of the Automobile, and then an Age of the Airplane.

The Age of the Railroad is perhaps the most romantic. For the automobile and the airplane have flourished in an already industrial nation, but the railroad often encountered a virgin continent. Railroads came on the American scene when Indians were still a menace to westering settlers, when large expanses of North America had not yet become accustomed to the English language or the European mode of dress. While railroads in Europe were commonly the servants of established communities, in America railroads were often their creators. For most of the nineteenth century railroads were a symbol of the anachronism which was the romance and the strength of the new nation.

But the Golden Age of the Railroads, as Mr. Stover vividly reminds us, is past. The peak in railroad mileage was reached in 1916. Since then the decline has been steady, and the process is not likely to be reversed. While that Golden Age is still a living memory, we might well try to recapture its spirit. In this brilliantly cogent story, Mr. Stover recalls that since their appearance on the American scene railroads have played a decisive role in nearly every major movement in our history. They were important in hastening the rise of the Atlantic seaboard metropolises, in peopling and supplying the West, in attracting and transporting immigrants, in shaping the enterprises of trappers, cowboys, miners, and farmers. In the twentieth century the commuting train has helped create and perpetuate the American suburbs. Railroads helped determine the course and the outcome of the Civil War, and were essential to victory in two world wars.

The story of railroads, as Mr. Stover tells it, is a parable of the changing American economy and the changing role of government in American life. Here we see how the American economy has or has not been competitive. We see how government subsidy, regulation, and control have encouraged the rise of railroads, have shaped their maturity, and have helped make their lives difficult, feasible, profitable, or impossible. Here we see also how the American economy has been integrated, homogenized, and brought increasingly under government supervision. In this brief volume, Mr. Stover gives us a much-needed factual guide to a subject which too often overwhelms the amateur; and he draws these facts into the drama of a rising American civilization.

By placing the history of American railroads in the main stream of American history, Mr. Stover admirably serves the purpose of the "Chicago History of American Civilization," which aims to make every aspect of our past a window to all our history. The series contains two kinds of books: a *chronological* group, which provides a coherent narrative of American history from its beginning to the present day, and a *topical* group, which deals with the history of varied and significant aspects of American life. This book is one of the topical group.

DANIEL J. BOORSTIN

Acknowledgments

I am indebted to Professor Richard C. Overton, of Manchester Depot, Vermont, whose excellent advice in the early planning of this volume was of the greatest assistance. Time for much of the writing was made possible by a research grant provided by the Purdue Research Foundation. An earlier grant from the same organization aided me in the research necessary for the first chapters of the book. Special thanks are due Miss Elizabeth Cullen of the Library of the Bureau of Railway Economics in Washington, D.C.; Mr. Thomas J. Sinclair of the Association of American Railroads; Mr. Roderick Craib, Associate Editor of *Railway Age;* and Dean M. B. Ogle, Jr., of Purdue University. The respective staffs of the Library of the University of Illinois and the Purdue University Library were generous with both time and assistance. Also I wish to thank Professor John H. Moriarty, director of the Purdue Library, for his counsel and advice. Finally, I wish to acknowledge the services of two competent secretaries, Mrs. Ruth Bessmer and Mrs. Kathryn McClellan.

In this second edition I am indebted to several people. Thomas Schmidt of CSX read the new chapters and made several pertinent suggestions to the text. Richard D. Meador, Jr., of Norfolk West-

Acknowledgments

ern, Gary Sease of CSX, Ken Longe of Union Pacific, Vicky Wells of the University of North Carolina Press, and Debra Basham of the West Virginia Division of Culture and History provided me with additional pictures for the book. I am most indebted to my wife, Marjorie, who typed both the text and the revised index.

I

"A Perfect System of Roads and Canals"

After the close of the war with England in 1815, Americans turned their attention from the Atlantic to domestic problems at home. The rich agricultural production of the country, the small but expanding factories of eastern cities, and the largely untapped natural resources of the nation—all of these called for improvements in transport. In the half-dozen years after the Treaty of Ghent nearly every state and large city started to agitate for an expanded system of internal improvements. Taverns and exchanges across the land heard the warm arguments of farmers, merchants, and politicians as they advanced the rival claims of canals and turnpikes, of steamboats and railroads. The domestic and western trade increased, even with the internal improvements still in the dream stage, and there was a relative decline in the importance of foreign trade. This was to be expected for the growing population, which was increasing roughly a third each decade, was on the march to the interior. The center of population, which in 1810 had been but a few miles up the Potomac from Washington, had moved by the eve of the Civil War to a spot west of Athens, Ohio. While in 1815 only four of the states had lacked a seacoast, the new states of the next half-century were nearly all in the interior.

The expansion and growth were more than sectional. As John C. Calhoun (1782–1850), at the time a nationalist leader, said in 1817: "We are greatly and rapidly—I was about to say fearfully—growing. This is our pride, and our danger; our weakness and our strength. . . . Let us, then, bind the Republic together with a perfect system of roads and canals." Even before the nation had its roads and canals, Oliver Evans (1755–1819), the frustrated Philadelphia inventor and steam engine builder, was having a bigger dream. In 1812 Evans foresaw the day when "carriages propelled by steam will be *in general use*, as well for the transportation of passengers as goods, travelling at the rate of fifteen miles an hour, or 300 miles per day." The country first had to build its turnpikes, dig its canals, and fill its rivers with steamboats before it could construct its railroads. But it was the dream of Oliver Evans that was to bring much of the material progress gained in the new century. Farmers, factory owners, and merchants were soon to become impatient with the interruptions and delays of the slow-moving wagons and boats that were typical even with completed turnpikes and canals. They did not like to wait upon late winters, spring freshets, or muddy roads. They desired a form of land transport that was fast, cheap, and dependable. By the standards of the early nineteenth century even the first railroads soon were to provide that kind of transportation.

With peace the nation quickly turned to internal improvements. Less than a month after the Battle of New Orleans, Colonel John Stevens (1749–1838) of Hoboken obtained from the New Jersey legislature the first railroad charter in America, a grant to build a railroad between the Delaware and Raritan rivers. In 1817, when the longest completed American canal extended less than 28 miles, the New York legislature authorized the building of the Erie Canal, a 364-mile project. The next year saw the completion of the Cumberland, or National, Road to Wheeling, Virginia, an event which made westward migration to the new states of Indiana and Illinois much easier. The average western emigrant in 1818 probably floated down the Ohio River from Wheeling in a flat-boat, but in that year at least a dozen small, wheezing steamboats were already plying western rivers, a number destined to increase tenfold in a decade. Late in 1825 the westward movement was stimulated by the completion of the Erie Canal. Much earlier in that year Colonel Stevens had confounded his critics, who were pointing to his chartered but unbuilt railroad, by operating the first locomotive to run on rails in America. His sixteen-foot "Steam Waggon" carried hardy house guests around circular track on the grounds of his Hoboken estate at the rate of 12 miles an hour.

For half a generation after Ghent the roads and canals desired by Calhoun made most of the significant transportation headlines in America. Much progress in the construction of improved toll roads had been made before the War of 1812. Generally built by private-stock companies with charters from the state legislatures, the turnpikes were built along the major travel routes. The companies were allowed to gain a return on their investment by charging tolls on the traffic using the new road. The early success of the well-built Lancaster Turnpike, completed from Philadelphia to Lancaster in 1794, resulted in a widespread demand for more roads. Dozens of turnpike companies planned and built roads in New England in the first years of the nineteenth century, and by 1825 a number of major roads criss-crossed southern New England. The record amount of land carriage caused by the blockade during the war with England made all sections of the country increase their road-building activity. The Pennsylvania legislature in 1816–17 granted state aid to 46 separate internal improvements, most of them turnpikes. By 1821 some 150 turnpike companies had been chartered in Pennsylvania, with state aid accounting for 35 per cent of the total subscribed capital of $6,401,000. Of the 1,807 miles of completed roadway, about two-thirds was well constructed with stone surfacing. In the same year New York could boast of 4,000 miles of completed roads, but the absence of any large amounts of state aid probably explained the lighter construction and lower cost or capitalization per mile in the Empire State ($3,500 per mile in Pennsylvania and $2,750 per mile in New York).

The rage for turnpikes was also present in the Old Northwest during the 1820s and 1830s, but few roads were ever completed, except in Ohio. In the southern states, except for Maryland, Virginia, and South Carolina, there was more promotion and planning than accomplished construction. Few of the turnpikes, North or South, ever achieved the profits promised by their promoters. New England was probably typical; only half a dozen of the 230 turnpike companies in the New England states ever returned their owners a reasonable dividend on investment.

The National Road was the greatest of all the turnpikes. Despite its extension to Wheeling in 1818 and to Columbus, Ohio, in 1833, this broad artery to the West had its problems. Although the federal government had spent $1,300,000 on the road in the preceding six years, Postmaster General Return J. Meigs, Jr. (1764–1824), complained to Congress in January, 1823, that on a recent inspection trip he found "the western (being the newest) part of the road . . . in a ruinous state, and being rapidly impaired." Along the route he saw rockslides and erosion so bad "that two carriages cannot pass each other."

Even with the advantage of turnpikes, freight rates were expensive. At the conclusion of the War of 1812 the lowest wagon freight rates averaged thirty cents per ton-mile and often were twice that. But farm produce was seldom carried overland. The reason was simple: transportation costs from Buffalo to New York City in 1817 were three times the market value of a bushel of wheat and six times that of corn. Wagon freight rates westward to Pittsburgh fell perhaps 50 per cent by 1822, but this was brought about as much by a general deflation as by an increase in the number of turnpikes or the growing competition of cheaper canal transportation. Nor was wagon freight fast transportation. When a Conestoga wagon drawn by four horses traveled the ninety miles from New York City to Philadelphia in three days, it became known as the "flying machine." Even though it was slow and costly, wagon traffic was heavy. In a two-and-a-half-day stage trip from Chambersburg to Pittsburgh in the fall of 1817, the Englishman Henry Bradshaw Fearon noted that his coach overtook 103 freight wagons headed for Pittsburgh.

Stagecoach traffic, too, was extensive. Sharing the road with the Conestoga wagon, the stagecoach, designed primarily as a passenger vehicle, offered a much faster means of transportation than the slow-moving freight wagon. The advantage of speed was offset, however, by the disadvantage of a rough ride. This was even true in the case of the Concord coach, which was far from comfortable. Of a trip through Pennsylvania, George Sumner (1817–63) wrote: "For two days and two nights was my body exposed to the thumps of this horrid road, and when I got to Pittsburgh (after having broken down *twice*, and got out *three* times during one night and broken down rail fences to pry the coach out of the mud) my body was a perfect *jelly*—without one sound spot upon it, too *tired* to stand, too *sore* to sit."

Both Albert Gallatin (1761–1849) in 1808 and John Calhoun a decade later stressed canals, as well as improved roads, in their plans for internal improvements. Canals are built slowly, and in 1817, when Calhoun made his plea for a perfect system of transportation, only about 100 miles of canals had been constructed, the longest single canal being just over 27 miles in length. But in that year Governor DeWitt Clinton (1769–1828) prodded the state of New York into starting a 364-mile canal from Albany to Buffalo, a project that would require $8,000,000, eight years, and unprecedented engineering feats. Completed in 1825, the Erie Canal was an immediate financial success. The canal craze was on.

By 1830 at least 10,000 miles of inland waterways were projected and 1,277 miles had been built. More than 2,000 additional miles were constructed during the thirties, and in 1840 the total canal investment in the nation stood at $125,000,000. Although in 1840 every state east of the Mississippi, with the exception of Illinois, Michigan, Mississippi, and Tennessee, had canals in operation, 70 per cent of the total mileage was to be found in Pennsylvania, Ohio, and New York. Since canals were much more expensive than turnpikes, generally costing from $20,000 to $30,000 per mile, most of the total investment in them, perhaps 60 per cent, was direct state aid. The Panic of 1837 revealed that several states had extended their credit too far in supporting internal improvements and were on the verge of bankruptcy.

In addition to the major waterways, several minor types of canals were built. These included short canals designed to avoid some river obstacle, such as that at the falls of the Ohio near Louisville; canals connecting two busy rivers, lakes, or bays, such as the Wabash and Erie in Indiana or the Chesapeake and Delaware in the East; and anthracite tidewater canals, such as the Lehigh or Morris canals in Pennsylvania. The most important type, however, was that intended to draw western business to some eastern city. Nearly every coastal city from Portland to Savannah had dreams of reaching the back country by canal, but frequently these dreams were thwarted by mountains or by lack of money. Only three westward-looking canals followed the Erie pattern: the "Pennsylvania System," the Chesapeake and Ohio, and the James River and Kanawha. The last two were stopped by mountains, the 184-mile Chesapeake and Ohio at Cumberland, Maryland, and the 200-mile James River and Kanawha at Buchanan, Virginia. The state of Pennsylvania eventually built its 395-mile Main Line to the Ohio River, but only with the assistance of 174 locks and several inclined-plane railroads (near Hollidaysburg) that rose 2,200 feet above sea level.

Whether canal projects brought profit or ruin to their sponsors, they all succeeded in greatly reducing the costs of freight carriage. Even on the difficult Pennsylvania Main Line rates were down to 8 cents a ton-mile in 1833, as contrasted to 20 cents by wagon. On the Erie, freight rates ranged from 1.6 cents to 3.4 cents per ton-mile in the thirties and forties and were down to a penny a mile by mid-century. Although the canal lineboat and the freighter took much business away from the Conestoga wagon, the stagecoach suffered less severely from the competition offered by canal packets or passenger boats. The packet could offer a smooth, leisurely trip, often with surprisingly good meals. Its five and a half by two foot bunks for overnight travelers were so closely spaced that Charles Dickens (1812–70) found it best to roll directly from the floor into his lower berth. For many the novelty of canal travel soon wore off. In 1830 Albany's five daily stage lines going westward up the Mohawk Valley often had to add extra coaches. The stage lines could also offer year-round service, while most canals were closed from three to five months each winter.

River transportation had always been vital in the United States, with its vast distances and extensive hinterland. By the early nineteenth century, eastern and western rivers alike were crowded with a variety of "bullboats," canoes, flatboats, and keelboats. Shortly before the War of 1812 the steamboat was added to this assortment. In 1811 the 371-ton Pittsburgh-built "New Orleans" successfully reached New Orleans, and four years later the "Enterprise" proved the feasibility of upstream travel with a round trip from Pittsburgh to New Orleans. By 1821 seventy-two boats were employed on western waters, and frequently a dozen could be seen together down in New Orleans. By 1830 the average steamboat was giving freight service faster than the canal boat and cheaper than the wagon.

For the passenger, steamboat accommodations were faster and more luxurious than those found on the canal boat and both cheaper and more comfortable than those of the stagecoach. Cabin passage, including meals, for the 2,064-mile trip from Pittsburgh to New Orleans cost only $45 in 1839. Most Americans could find little fault with the luxuriously appointed craft and were proud when European visitors like Count Francesco Arese admitted that "the finest ones we have in Europe are much inferior to the smallest, the wretchedest ferry-boat over here." Steamboat travel offered not only luxury but also danger. Old wrecks lined every river by mid-century, and the newspapers were filled with stories of explosions, collisions, fires, and snagged boats. Accidents on western rivers in 1853 claimed 78 boats and 454 lives. The optimistic traveler could take satisfaction in the claim of *Niles' Weekly Register* that "with moderation and care steamboats are not so very dangerous." River travel was less dangerous in the East, and on the Hudson River a five-year experiment with safety barges (passenger boats towed behind steamboats) ended in failure in 1830 when the barges "Lady Clinton" and "Lady Rensselaer" were offered for sale as freight towboats. But even on eastern rivers the natural factors of ice and low water could halt all navigation. Between 1818 and 1859 ice closed the Hudson at Albany an average of four months a year, and at Pittsburgh ice, plus seasons of low water, made a comparable reduction in commerce.

In the first decades of the nineteenth century many of the needs for better transportation were fulfilled as well-constructed roads, a host of new canals, and faster steamboat travel appeared. In the years after the War of 1812 nearly every ambitious American city was the center of a radiating network of turnpikes, either planned or already built. In 1826 Boston had eighty stage lines with more than two hundred scheduled arrivals and departures each day. The toasts, processions, transparencies (illuminated banners), and kegs of river water from around the world that were employed to mark the opening of Governor Clinton's Erie Canal in the autumn of 1825 initiated a series of canal celebrations which continued for years. High- and low-pressure steamboats brought prosperity to

Pittsburgh, Cincinnati, Louisville, and St. Louis and dreams of the same to the remote upriver towns of Knoxville and Indianapolis.

Wagon freight never became really cheap, and the best Concord coaches were still slow and uncomfortable. Canals were stymied by mountains and were certain to freeze over in winter. Steamboats, although they were fast and luxurious, were also dangerous. Moreover, Henry Miller Shreve (1785–1851), veteran steamboater and inventor of the "snagboat," which became known as "Uncle Sam's tooth-puller," could not always remove hazardous obstructions from river channels, even for the record shallow-draft "Orphan Boy" (169 tons, 22-inch draft). Some new form of transportation, year round in regularity, safe and cheap, overland and unlimited in route, was obviously needed. Americans in the 1820s did not have long to wait.

2

First Rails

The iron rail, flanged wheel, and puffing locomotive appeared in America by 1830. In the next twenty years the railroad brought a new dimension and added a new flavor to American transportation. The first railroads frequently helped American cities (and in turn were aided themselves) as they sought a larger share of western markets.

In the first third of the nineteenth century the commercial rivalry among major eastern American cities was as intense as the competition among the various transportation facilities that served them and the nation. Already first in population by 1810, New York City increased her lead after 1815 through a combination of natural geographic advantages and a series of aggressive measures pushed by an energetic citizenry. Encouraged when the British chose New York as a market for surplus British goods after the War of 1812, the merchants of that city increased this trade advantage in 1817 by creating an attractive auction system for the sale of imports and by establishing dependable and regular packet service across the Atlantic.

Within fifteen years the expanding foreign trade of New York City was nearly half the total foreign trade of the nation, and by

1828 the New York Custom House was collecting sufficient duties to pay the entire expenses of the federal government. In the same period, shipping interests in the city, seeing the advantage of attracting the immigrant trade, were so successful that by the thirties roughly half the newcomers arriving in America were landing at New York. Simultaneously, the New Yorkers succeeded in securing their full share of the nation's domestic trade by developing extensive coastal commerce along the Atlantic and by catering to western trade after the completion of the Erie Canal in 1825.

The immediate success of Governor Clinton's favorite canal stirred rival cities from their lethargy. In the decade after 1825 several cities (Philadelphia with a canal, Boston and Charleston with railroads, and Baltimore with both) sought western trade. Philadelphia started first. Alarmed by the drop (within a generation) from first city to fourth in the volume of foreign trade, the city fathers vigorously supported the Pennsylvania state project to reach Pittsburgh by a combined canal and rail route. The 395-mile Main Line from Philadelphia to the Ohio was built between 1826 and 1834 at a cost nearly twice that of the Erie and across mountains nearly four times as high as those along the lower New York route. The project never prospered. In the first decade of its operation the revenues did little more than meet the current operating expenses of the system. The lack of success by the merchants of Philadelphia lay in their failure to appreciate the height of Pennsylvania mountains and the relative merits of a new form of transportation, the railroad.

To the south, Baltimore was somewhat more successful in her bid for western trade. This small town of colonial times had grown more rapidly between 1790 and 1820 than any other major American city. Ahead of Boston and Charleston and third in population by 1800, Baltimore boasted 35,000 people in 1810 and 63,000 in 1820. Her prosperity in the first two decades of the century was aided by the construction of the Cumberland Road and by the concentration of Baltimore merchants on both Susquehanna Valley trade and the coastal commerce built up by fast "clipper" schooners. Well aware of their own commercial progress, the citizens of Baltimore could generously agree with their local editor, Hezekiah Niles (1777–1839), when he wrote of New York City in 1821: "This [city] is now the second commercial city in the world; a little while and it will probably be the *first*, by means of its canals and the trade of the lakes." Baltimore and Maryland were a little slower than Philadelphia and Pennsylvania in meeting the challenge of New York's canals, but when they did move, they made a dual attack. Both the state of Maryland and the city of Baltimore were sanguine about the success of their canal and railroad projects, since they could claim a shorter route to the Ohio, by a margin of two hundred miles, than either of the rival northern cities.

At Georgetown on July 4, 1828, John Quincy Adams (1767–1848) wielded a ribbon-bedecked spade to start the Chesapeake and Ohio Canal; forty miles away, near Baltimore, the aged Charles Carroll (1737–1832) was laying the first stone for the Baltimore & Ohio Rail Road. The Baltimore people soon realized that the canal would favor the new national capital far more than their own city and therefore transferred their full allegiance to the railroad. The plump and pleasant president of the new railroad, merchant-banker Philip E. Thomas (1776–1861), pushed construction with vigor, and by May 1830, passengers could enjoy a thirteen-mile trip by horse-drawn car out to Ellicott's Mills, Maryland.

Steam power came to the Baltimore & Ohio later that summer when Peter Cooper (1791–1883), glue-maker from New York City and part-time inventor who had just taken a flier in Baltimore real estate, built the experimental locomotive "Tom Thumb." The usefulness of the little engine was limited, however, and except for its famous race with a horse and occasional trips for the benefit of distinguished guests, it stood idle. The railroad used horse-drawn cars until extensive trials in the summer of 1831 proved the three-and-one-half-ton "York" to be a fully practical and serviceable locomotive. By the time of his retirement in 1836, Thomas had extended the railroad to Harper's Ferry and had built a 37-mile branch line to Washington. The railroad now had gross revenues of over $260,000 yearly and claimed 7 locomotives, 1,078 "burthern," or freight cars, and 44 passenger cars. The new president, Louis McLane (1786–1857), former Jacksonian cabinet member, pushed the B.&O. on to Cumberland by 1842, but further extension to the west was slowed by high mountains and opposition from both Virginia and Pennsylvania. Not until Christmas Eve, 1852, was the last spike driven in the line which reached the Ohio River at Wheeling, Virginia. By this time the merchants in New York City were less inclined to worry about their rivals in Maryland.

In the South, Charleston, a city whose foreign trade had gone into decline, bid for a greater share of the inland trade by building a railroad. The merchants of the city, hoping to secure the trade of a rich cotton-growing area in their own state and in Georgia, projected a railway to Hamburg, South Carolina, just across the Savannah River from Augusta, Georgia. As the first rails were being laid out of Charleston by the South Carolina Canal and Rail-Road Company, the new chief engineer, Horatio Allen (1802–90) of New York, persuaded the directors to use steam power on the line. "There is no reason to expect any material improvement in the breed of horses in the future," argued Allen, "while, in my judgment, the man is not living who knows what the breed of locomotives is to place at command." On Christmas Day, 1830, the "Best Friend of Charleston," constructed for $4,000—the first locomotive built for sale in the United States—carried 141 passengers on the first scheduled steam-railroad train run in America. The entire 136-mile route to Hamburg was completed by October 1833, to become the longest continuous railroad in the world. In later years

South Carolina built additional mileage more slowly, for the total remained at 136 miles in 1840 and had only doubled by 1850.

In New England, Boston was worried about the commercial success of New York. Not only did the Erie Canal threaten to capture Boston's trade with western Massachusetts, but much business in the central section of the state was being attracted by New Haven, Hartford, and Providence. In their quest for a commercial revival, the Boston merchants wisely rejected the idea of more canals—waterways which, after all, were expensive to build and which, because of winter ice, were at best open only two-thirds of the year. Bostonians noted the success of the Granite Railway, a two-mile broad-gauge tramway built at Quincy by Gridley Bryant (1789–1867) in 1826 to transport granite for the Bunker Hill monument, and the following year (May 1827) they eagerly paid admission to see the display of English locomotives at the nation's first Railway Exhibition. After efforts to obtain state financial support for a western railroad from Boston to Albany failed, three short, privately financed lines were chartered in 1830–31. To the north the Boston and Lowell linked two important cities. To the south the Boston and Providence provided a shortcut over a longer water route. The third line, the Boston and Worcester, was seeking the western trade. All three lines were in operation by 1835 and they were to form the nucleus of the New England rail system.

Even before the completion of the Boston and Worcester, its president, Nathan Hale (1784–1863), long-time editor of the *Boston Daily Advertiser* and nephew of the Revolutionary War patriot, was urging that the rail line be continued to the west. The Western Railroad, to run the 150 miles from Worcester, via Springfield and Pittsfield, to Albany, was chartered in 1833, and by the end of 1835 all of its stock had been sold, most of it in Boston. The $2,000,000 in private money was admittedly inadequate for the ambitious project, and the Commonwealth soon subscribed $1,000,000 in stock and by 1841 had lent an additional $4,000,000 to the line.

The road was completed to Springfield by 1839 and in the next two years was pushed through the rugged Berkshires to the Hudson River and Albany. The city fathers of both Boston and Albany made the first round trip on the line, the Boston delegation returning home on December 28, 1841, with a barrel of Rochester flour, which was used to bake bread for that evening's celebration. Boston now possessed an all-rail connection with the west and the Erie Canal, but she never obtained any major portion of the traffic destined for New York City via the Hudson River. After all, the Western Railroad, to use the language of Charles F. Adams, Jr. (1835–1915), was built upon "the fallacy that steam could run up hill cheaper than water could run down." But in completing this early transectional railroad (later consolidated as the Boston and Albany), Boston had built a line noted for its financial sobriety and technical excellence.

Few railroads were projected or built without some opposition, often a considerable opposition, from a variety of timid citizens, vested interests, tavern keepers, turnpike and bridge companies, stagecoach lines, and canals. The projected rail route from Boston to Albany was opposed by a Boston editor who thought the project not only impracticable but also "as useless as a railroad to the moon." Even when completed the railroad frequently had to overcome the opposition of state governments loyal to their canals. In New York railroads running parallel to the Erie Canal had to pay tolls equal to those of the canal, and in Ohio and Pennsylvania special taxes were levied against rail traffic which competed directly with canal business. In Ohio a school board placed a moral judgment on rail lines, calling them "a device of Satan to lead immortal souls to hell." When members of the Massachusetts legislature chartered the state's first railroad, they were described by turnpike interests as "cruel turnpike killers and despisers of horseflesh." The wealthy Philadelphia merchant Samuel Breck complained that "if one could stop when one wanted, and if one were not locked up in a box with 50 or 60 tobacco-chewers; and the engine and fire did not burn holes in one's clothes . . . and the smell of the smoke, of the oil, and of the chimney did not poison one . . . and [one] were not in danger of being blown sky-high or knocked off the rails—it would be the perfection of travelling."

For most Americans, however, the decade of the thirties was an era of railroad enthusiasm and noisy railroad fever. Within half a dozen years after the Baltimore & Ohio began operations, some two hundred lines had been chartered, taking the savings of thousands of investors, many of whom had never seen a railroad track or a steam locomotive. Those with pennies instead of dollars joined the rail craze by purchasing glassware, china, sheet music, or wallpaper decorated with brilliant pictures of railroad cars and steaming locomotives. The young nation dreamed and planned ambitious rail lines that were to cross unsettled territory, span rivers, and reach distant cities.

Early in 1830 the two sons of Colonel John Stevens—Robert Livingston Stevens (1787–1856) and Edwin A. Stevens (1795–1868)—obtained a charter from the New Jersey legislature for the Camden and Amboy Railroad and Transportation Company. Written into the charter was a "monopoly" clause which gave the Stevens boys exclusive rail-transportation rights between New York City and Philadelphia. This advantage assured a rapid subscription to the one-million-dollar stock issue, and the road was completed across the state by 1833. The English-built locomotive "John Bull" provided a seven-hour passenger service between the two cities for a three-dollar fare. Connecting service westward from Philadelphia to the system of state canals was available in 1834 on the state-owned Philadelphia and Columbia. Four years later a main line to Baltimore was opened with the completion of the Philadelphia, Wilmington, and Baltimore.

The first railroad in New York State was conceived when English-born George William Featherstonhaugh (1780–1866) ob-

"The First Railway Train" by E. L. Henry re-creates the first trip of the "DeWitt Clinton" between Albany and Schenectady in 1831. (Courtesy, Albany Institute of History and Art.)

Chicago's first railroad station. When this depot was built in 1848 by the Galena and Chicago Union Railroad, William Ogden or his agents spotted incoming trains from the cupola. The picture was taken somewhat later, since such a fringe benefit as a "Railway Mens Reading Room" was not common in the middle of the nineteenth century. (Courtesy, Chicago & North Western Railway.)

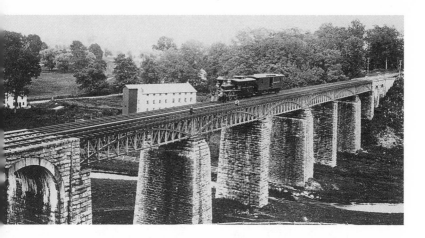

Valley Creek Bridge, Coatesville, Pennsylvania, in 1860. This well-built bridge reveals a fundamental difference between the railroad and the highway. A road would probably have used a ford to cross this small creek. (Courtesy, *Railway Age*.)

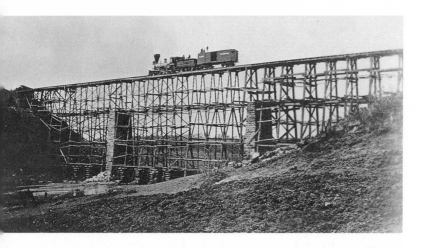

Potomac Creek Bridge, constructed in May 1862 in nine working days by unskilled soldiers of the U.S. Military Railroads under the supervision of General Herman Haupt. Even after seeing this bridge carry heavy loads, Abraham Lincoln described it as apparently constructed of nothing more than "beanpoles and cornstalks." (Courtesy, Association of American Railroads.)

tained a charter for the Mohawk and Hudson Rail Road. This seventeen-mile line between Albany and Schenectady would permit canal-packet passengers to avoid the day-long forty-mile ride through the many locks in the terminal section of the Erie. The road was built in 1830–31 by John B. Jervis (1795–1885), recently of the Delaware and Hudson Canal and Railroad Company. Service began in August, 1831, with the company's first locomotive impudently named the "DeWitt Clinton." Less than six months later *Niles' Weekly Register* reported that twenty-five different New York railroads (with capital structures amounting to $41,000,000) were applying for incorporation before the state legislature.

The railroad industry expanded rapidly during the thirties as nearly every state laid rails. Of the twenty-six states in 1840, only four (Arkansas, Missouri, Tennessee, and Vermont) had not completed their first mile of track. Most of the lines were built to serve coastal cities in the eastern states, for only a twelfth of the 3,000 miles of road lay west of the Appalachians. The New England and Middle Atlantic states accounted for more than 60 per cent of the mileage in 1840, with Pennsylvania first in the nation in mileage (754), New York second (374), and Massachusetts third (301). Southern states could claim 1,105 miles of road in 1840, while the Old Northwest, though dozens of lines were projected there, completed only 133 miles in the thirties.

The Panic of 1837 and the depression which followed were disastrous for the rail projects of some states, especially Michigan, Indiana, and Ohio, but for most of the nation no immediate slackening of construction was apparent, since about half the decade's new lines were built in the last three years—1838, 1839, and 1840. While additional canal construction was carried on during the thirties, it never quite matched railroad-building, and thus in 1840 the two facilities were nearly equal in mileage. The western states pushed their canal projects vigorously during the decade and accounted for nearly a third of the nation's 3,300 miles of canals, with

Ohio (744 miles) being second only to Pennsylvania. Although American railroads were generally following English technique and experience during the thirties, the United States easily outstripped total European rail construction during the decade by a margin of 3,000 miles of new road in America to only 1,800 miles in Europe. Although western Europe possessed the advantages of better engineering, more highly developed metal-working techniques, and easier financing, the United States more than balanced these with the assets of a greater need for improved transport, a relative freedom from long-entrenched customs, prejudices, and monopolies, and cheaper land for railroad right of way. America had about $75,000,000 invested in her rail network in 1840, which was far less money than the European expenditure for a much smaller system.

In 1840 the American railroad system was a thin, broken network stretching along the Atlantic coast from Portsmouth, New Hampshire, to the Carolinas. If a traveler wanted to go from Boston to Georgia, he could go by rail to Stonington, Connecticut; by steamer to New York City and across to Jersey; by rail to Washington, via Camden, Philadelphia, and Baltimore; and by steamer forty miles down the Potomac to rail service which would get him to Wilmington, North Carolina. After a stage trip through South Carolina to Charleston, he could continue by train on the South Carolina Railroad to the end of the track at Hamburg, across the river from Augusta, Georgia.

Technical innovations and improvements occupied the minds of all the early railroad-builders. The pages of the first important railroad periodical, the *American Railroad Journal*, founded by D. Kimball Minor in 1831, were filled with technical and engineering articles throughout the thirties and forties. Railroad companies quickly demonstrated the necessity of their controlling the traffic on the lines they built. Thus, unlike the canal and turnpike companies, the railroads soon universally owned and operated all the rolling stock and motive power on the road. As "common carriers," they were expected to accept for shipment anything within reason.

Some of the first railroad improvements were focused on the roadbed. Nearly all of the early railways had built their track with long iron straps, or bars, fastened to wooden rails, which, in turn, were secured to large blocks of granite or other stone. The iron-strap rails, although twenty to twenty-five feet in length, had a tendency to work loose and curl up under the weight of trains to form "snakeheads," which often broke through the floor of a passing coach. The strap rails could be increased in thickness, but the real solution to the problem appeared when Robert L. Stevens, president and engineer of the Camden and Amboy Railroad, designed the T-rail. Stevens carved his first model rail out of wood while in England on a locomotive buying trip, had the rails made in Wales, and was installing them as original track on his New Jersey line in 1831. The new type of rail was not only strong but its flat bottom (requiring no rail chair as was common in England) could be spiked directly to the block or tie. With the increasing weight of locomotives and rail traffic, the T-rail soon became standard equipment on all first-class lines. The state of New York required the new type rail on all railroads after 1847. On the branch line or the cheaply constructed road, strap-iron rails, frequently second hand, remained a common sight until the Civil War. The heavy stone or granite blocks also proved unsatisfactory, both because of their susceptibility to frost-heaving and also because the track proved so unresilient as to be hard on the rolling stock. Wooden crossties imbedded in a gravel roadbed were found to furnish the most satisfactory type of rail construction. Railroad managers soon realized that even the best-laid original track could not be considered permanent and that the costs of maintaining track and roadbed would be continuous.

Early American railroad lines presented a great variety in

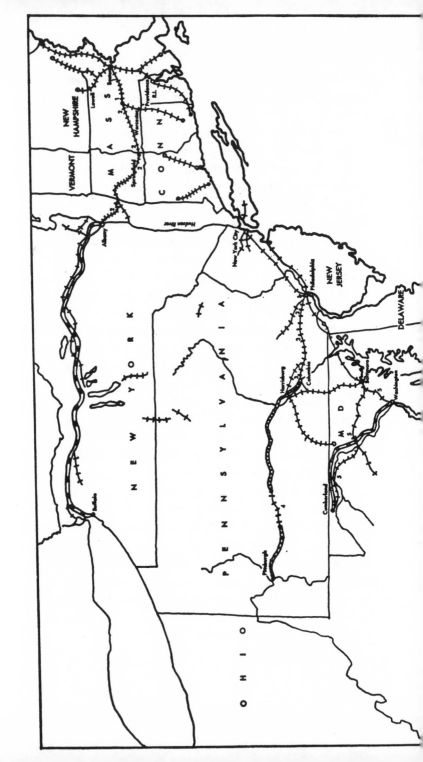

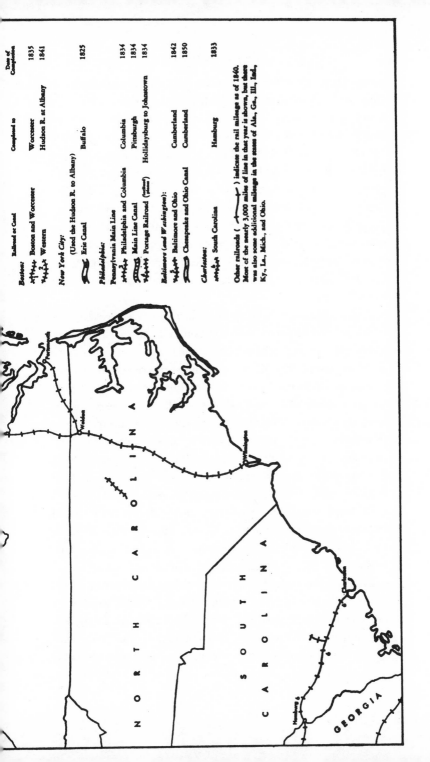

Railroad or Canal	Completed to	Date of Completion
Boston:		
Boston and Worcester	Worcester	1835
Western	Hudson R. at Albany	1841
New York City:		
(Used the Hudson R. to Albany)		
Erie Canal	Buffalo	1825
Philadelphia:		
Pennsylvania Main Line		
Philadelphia and Columbia	Columbia	1834
Main Line Canal	Pittsburgh	1834
Portage Railroad	Hollidaysburg to Johnstown	1834
Baltimore (and Washington):		
Baltimore and Ohio	Cumberland	1842
Chesapeake and Ohio Canal	Cumberland	1850
Charleston:		
South Carolina	Hamburg	1833

Other railroads (————) indicate the rail mileage as of 1840. Most of the nearly 3,000 miles of line in that year is shown, but there was also some additional mileage in the states of Ala., Ga., Ill., Ind., Ky., La., Mich., and Ohio.

gauge. Many of the first roads, since they used English-built engines, naturally built to the English gauge of 4 feet 8½ inches, a figure long customary to English wagons. This was the most common gauge in New England and in the North, although much variation was to be found in Ohio and Pennsylvania. In New York the Erie Railroad deliberately—and unwisely—used a broad gauge of 6 feet to prevent loss of traffic to other lines. This variation proved costly, and eventually the Erie shifted to the standard gauge. In most of the southern states the 5-foot gauge was common, though certainly not universal. As early as 1834 the *American Railroad Journal* was stressing the necessity of uniformity in gauge, but this ideal was not to be achieved on a truly national basis for half a century.

A more rapid development of uniformity was achieved in motive power. While horses, mules, stationary engines, and even sails and horse-operated treadmills were tried on the early lines, the steam locomotive soon relegated the horse to a reserve and substitute status. Within a few years most railroad horses were back on the farm. Some of the first American-built engines were not entirely satisfactory. When Philadelphia jewelry manufacturer Matthias Baldwin (1795–1866) built his first full-size locomotive, "Old Ironsides," for the newly incorporated Philadelphia, Germantown and Norristown Railroad, the engine averaged only a mile an hour in its first trials. However, experimentation soon pushed the speed of Baldwin's locomotive up to twenty-eight miles per hour. One of the most important improvements was made by John B. Jervis while he was still chief engineer on the Mohawk and Hudson. Dissatisfied with the rigid front axle and poor turning characteristics of the heavy English engines, Jervis, with some advice from Horatio Allen, designed the first swivel, or "bogie," wheels, which were mounted in a truck under the front of the locomotive. Jervis' new locomotive, the "Experiment," built in 1832 by the West Point Foundry, negotiated all curves with ease and soon proved capable of speeds approaching a mile a minute. The equalizing beam, invented by Joseph Harrison (1810–74) of Philadelphia in 1839, was also important because it permitted equal pressure by each drive wheel, even on rough or uneven track. Even earlier, however, in 1836, Henry R. Campbell (1781–1844) of the same city had designed an eight-wheeled engine (bogie truck plus four drivers) which, as the American-type locomotive, was to dominate American locomotive styling for half a century. James Brooks built the first engine of this type in 1837.

In the thirties the locomotive acquired many of the features which made its silhouette distinctive. Robert Stevens and his top mechanic, Isaac Dripps (1810–92), probably invented the first "pilot," or cowcatcher, for the locomotives on the Camden and Amboy. When early trains collided with a cow, the engine often landed in a ditch or in the repair shop, but after Isaac Dripps perfected the cowcatcher, the railroad company had to pay for the cow. In South Carolina Horatio Allen initiated the first train travel at night when he built a fire of pine knots in sand on a small flatcar pushed ahead of a Charleston and Hamburg locomotive. Night trips on the railroad lines of the thirties were very infrequent, and thus the conventional headlight, burning kerosene in front of tin reflectors, became common only in the forties. The sandbox (to provide improved wheel traction) first appeared in Pennsylvania in 1836 when a great plague of grasshoppers visited the state and threatened to bring all train movements to a halt. Later that same year, George W. Whistler (1800–49), former army lieutenant, surveyor of the Baltimore & Ohio, and husband of the lady portrayed in "Whistler's Mother," built the locomotive "Susquehanna," which boasted the first steam-locomotive whistle in America. First used for signaling the brakeman and the train crew and for general warnings to a more pedestrian traffic, in later years the whistle became a sort of occupational signature for the engineer who was skillful with the whistle cord. The locomotive cab was first tried on

Massachusetts lines as they were extended into New Hampshire. Canvas (and, later, wooden) cabs kept the engine crews from freezing to death in the bitter New England winters.

In the forties American rail construction easily kept pace with technical advances. The network of 3,000 miles of line in 1840 had tripled to nearly 9,000 miles of line in 1850, with every state east of the Mississippi claiming at least some mileage. Although railroads were being built in all sections of the East, the New England and Middle Atlantic states still had a clear lead at mid-century with more than 5,300 miles of railroad tracks, or nearly 60 percent of the national figure.

The rapid northeastern rail expansion was most noticeable in New England, where the forties saw a fivefold increase in mileage. In 1850 the railroad map of New England looked fairly complete, with three states (Massachusetts, Connecticut, and New Hampshire) already having mileage equal to roughly half of what they would have one hundred years later. Boston was the hub of the New England system, with lines radiating inland in every direction. A second major Massachusetts line, projected across the northern part of the state toward Troy, on the Hudson, was two-thirds completed by 1850. The expansion of the New England network could probably be traced to the fact that in the early thirties Massachusetts and Boston had chosen the railroad as the transportation of the future while rival states and cities to the south were still flirting with the canal. If some New Englanders complained in the forties that their region was the only one overbuilt with railroads, at least it resulted in New England's becoming the mother of railroading and railroad men for half a century.

In the same decade the railroads of New York and Pennsylvania also grew. In New York the connecting road west of the Mohawk and Hudson, the 78-mile Utica and Schenectady, was so prosperous that in 1850 the stockholders of the line rewarded their president, Erastus Corning (1794–1872), with a $6,000 service of plate. This same Erastus Corning, nail-maker and long-time mayor of Albany, was to lead in the 1853 creation of the New York Central Railroad, a consolidation of the ten little railroads strung along the Mohawk Valley and the Erie Canal. In 1850 two rival roads on the eastern side of the Hudson, the New York and Hudson and the New York and Harlem, had nearly completed their lines upriver to Albany. And in the state's southern tier of counties, the president of the New York and Erie Railroad, Benjamin Loder (1801–76), was driving his Irish and German construction crews to complete, by 1851, his 483-mile route to Dunkirk on Lake Erie. With such a variety of big rail-construction projects, it was easy for New York to claim first place (1,361 miles) among the states in 1850.

Pennsylvania was a strong second in mileage at mid-century. The major new line in Pennsylvania in the forties was the Pennsylvania Railroad, incorporated by Philadelphia businessmen in April 1846. Realizing the futility of depending upon the state canal system and seeing the commercial menace of the Erie Railroad taking shape to the north and the Baltimore & Ohio to the south, the merchants of Philadelphia decided to build a line to Pittsburgh. Under the strong, sure guidance of their chief engineer, John Edgar Thomson (1808–74), they completed, by 1850, a 137-mile road west from Harrisburg to connect with the state-owned inclined-plane railroad near Hollidaysburg.

Much new mileage was constructed in southern and western states. Every southern state added railroads in the forties, but Georgia far outstripped her neighbors, building 450 miles during the decade to achieve a position of railroad leadership in the Southeast that she was to maintain for a century. By 1850 Atlanta was the center of a growing network of lines. To the east the well-established Georgia Railroad, built by John Edgar Thomson before he returned to Pennsylvania, gave service to Augusta; to the southeast the Macon and Western Railroad and the Central Railroad reached the coast at Savannah; and to the northeast the 138-mile state-built Western and Atlantic had nearly reached Chattanooga.

Most of the 1,276 miles of the line in the Old Northwest in 1850 was located in Ohio (575 miles) and Michigan (342 miles). In Ohio a broken complex of lines stretched southward from Cleveland and Sandusky on Lake Erie to Columbus and Dayton and then on to Cincinnati on the Ohio River. All the mileage in Michigan was in the southern part of the state, with the major line, the Michigan Central, once a state-owned line but now owned by men from Boston, connecting Detroit and Lake Michigan.

The 9,000-mile national rail network of 1850 had cost perhaps $310,000,000 to construct, nearly twice the $160,000,000 sum invested in the 3,700 miles of canals at mid-century. The national average cost of railroad construction was $34,000 per mile, but there was a great variance in this figure in the several sections of the country. In New England, where the terrain was often rugged and the traffic fairly heavy, railroads had cost an average of $39,000 per mile, with the solidly built Boston and Lowell requiring $71,000 per mile, the Western Railroad costing $50,000, and the shorter lines, the Old Colony and the Fitchburg, costing only $25,000 and $30,000, respectively. In the Middle Atlantic states high construction costs (averaging $46,000 per mile) were caused as frequently by the unevenness of the financial management as by the roughness of the terrain. Upon completion in 1851 the Erie was valued at $43,000 per mile, exclusive of equipment. Erastus Corning's Utica and Schenectady was listed at $53,000 per mile in the same year and the Baltimore & Ohio at even more—$54,000 per mile. In southern and western states, where the terrain was fairly even and where the potential traffic volume often permitted a lighter type of construction, the average cost per mile was no more than $21,000, with the well-built Georgia Railroad costing only $17,000 a mile.

Over 50 per cent of the more than $300,000,000 invested in American railroads in 1850 was represented by capital stock, the remainder being in bonds. Since their required capital investments were always large and carried maximum risk, nearly all railroads organized as corporations, securing their charters from the state legislatures. These charters frequently included such liberal privileges as monopoly provisions (as in the case of the Camden and Amboy), partial or temporary exemption from state taxation, lottery or banking rights (notably the lines in Georgia), and, of course, sweeping privileges of eminent domain. The privilege of eminent domain not only permitted a railroad real freedom in the selection of its route, but also tended to keep its right-of-way costs low (unlike in England, where the railways had no such privilege). Since several state governments had, in earlier years, sponsored and built canals, it was natural that some of the first railroads were also state projects. In Pennsylvania two of the earliest lines in the state, the Portage Railroad and the Philadelphia and Columbia, were constructed with state money, as was the strategically located Western and Atlantic in Georgia. Three western states, Michigan, Indiana, and Illinois, also experimented, with very indifferent success, with public construction.

Infinitely more important than state construction was state financial assistance to privately owned lines. Sometimes, especially in the case of Virginia, this aid was granted through the purchase of capital stock, but normally it was in the form of loans. State financial aid to railways had already reached nearly $40,000,000 by 1838, and it was to increase still more by mid-century. The Western Railroad in Massachusetts, the Erie in New York, the Baltimore & Ohio in Maryland, and most of the railroads in Virginia were among the rail recipients of state assistance. There was little municipal aid in the thirties and forties, but for a few lines it was vital. Federal aid, prior to the 1850 land grant to the Illinois Central–Mobile and Ohio project, consisted of a number of route surveys by army engineers and a reduction in the tariff on imported railroad iron in the years from 1830 to 1843.

Most of the money for the early railroads came from private investors. Farmers and tradesmen living along the route of the proposed railroad, merchants, professional men, and businessmen living in the terminal cities—all were persuaded to subscribe to capital stock or buy bonds. The hope of improved land values, more profitable markets, or a general rise in prosperity was frequently in the mind of the investor, as was the thought that the investment itself might be a good one. Too often the early enthusiasm created in mass meetings or railroad conventions melted away when the subscription books were opened. Press and pulpit, pamphleteer and orator often were called upon to make the filled and signed subscription book a reality. The selling job was easier in those large eastern seaports (or important trading centers of the interior) where the railroad was becoming a major weapon in the competitive municipal mercantilism of the period. In the years before 1850 only minor portions of the total rail investment came from eastern banks or the money markets of Europe.

Passenger traffic and the revenue it brought in were much more important to the railroads of a century ago than they are today. The novelty of travel by rail plus the appeal of rocketing downgrade at perhaps 20 miles per hour helped furnish the first substantial financial returns for many new lines. Some of these roads, such as the Utica and Schenectady in 1836, were used exclusively for passenger business in the first weeks and months. In its fourth year of operation (1833–34) the Baltimore & Ohio carried 95,000 passengers with a revenue amounting to well over 40 per cent of the total for the year. A few lines, like the Syracuse and Utica, where the Erie Canal took the cream of the freight traffic, remained primarily passenger roads until mid-century. For the typical railroad, however, passenger traffic probably produced between a fourth and a third of the total revenue, a ratio which was common until after the Civil War. Early railroad passenger fares were frequently high, but by the late forties competition had reduced them to perhaps 2.5 to 3.5 cents per mile in New England and in the Middle Atlantic States and no more than 4 to 5 cents in the rest of the nation.

The first railroad passenger cars slavishly followed the stagecoach in design, but when Andrew Jackson (1767–1845) became the first president to travel by train, on the Baltimore & Ohio in 1833, the first corridor-type coaches had appeared. Ross Winans (1796–1877) came to Baltimore in 1828 to sell horses to the new B.&O. but stayed on to become an engineer on the line, inventing the long coach with trucks at either end of the car and the famous "camel-back" locomotive. Fanny Kemble (1809–93), the English actress, left this vivid description of a coach used on her American tour: "The windows . . . form the walls on each side of the carriage, which looks like a long greenhouse upon wheels; the seats, which each contain two persons (a pretty tight fit, too), are placed down the whole length of the vehicle, one behind the other, leaving a species of aisle in the middle for the uneasy (a large portion of the traveling community here) to fidget up and down, for the tobacco-chewers to spit in, and for a whole tribe of itinerant fruit and cake-sellers to rush through, distributing their wares at every place where the train stops." Miss Kemble may well have ridden in the Erie cars, with their water boys—youngsters who passed through the train with a long-spout can and a couple of glasses, complete with germs, for the benefit of thirsty travelers.

Coach travel was rarely comfortable. The cars were inadequately heated in winter and hot in summer. The menace of sparks kept the windows shut, although the intrepid Davy Crockett (1786–1836) insisted on opening one, claiming that "I can only judge of the speed by putting my head out to spit, which I did, and overtook it so quick, that it hit me smack in the face." Crude sleeping cars were introduced, with indifferent success, on several eastern roads in the thirties and forties, and the Camden and Amboy offered de luxe service in 1840 when it outfitted a pair of coaches with rocking chairs. The speed of the Camden and Amboy and other eastern trains appealed to the Post Office Department, and by 1834 Postmaster General William T. Barry (1785–1835) had contracted with several of them to help carry the "Great Eastern Mail." By the end of the decade William F. Harnden (1812–45) and Alvin Adams (1804–77) were renting space in baggage cars for their railway-express business.

In mid-century the freight business of American railroads was less exciting than the passenger traffic but economically more important. Even the first railroad freight rates were substantially lower than those of the wagon companies. In 1832 the *American Railroad Journal* reported that the proprietor of the mills at Ellicott's Mills, thirteen miles out of Baltimore, was receiving good freight service from the Baltimore & Ohio Railroad for only a fourth of his former transportation costs. In the following year the ton-mile rate on the new Boston and Worcester line in Massachusetts was only a third of the wagon or turnpike charges. Especially where traffic was heavy or distances great, the superiority of the railroad over wagon and turnpike freighting was obvious.

The railroad was slower in showing its supremacy, in freight carriage, over the canal as few rail lines could come close to canal transportation rates, especially for heavy, bulky goods. In 1853, when the ton-mile rate on the Erie and the Ohio canals was averaging just over a penny, the comparable rates on the New York Central, the Erie, and the Pennsylvania railroads ranged from 2.4 cents to 3.5 cents. For the less efficient or the poorly located canal, defeat by the railroad came more quickly. While the Erie Canal was still expanding its business in the fifties, its southern rival, the Pennsylvania Main Line Canal, ceased important operations shortly after the Pennsylvania Railroad reached Pittsburgh in 1852. Interruptions in service caused by winter ice, too much water, or too little water plagued all the canals. As early as 1831 *Niles' Weekly Register* reported that while both the Erie and the Delaware and Hudson canals had been closed for five months during the year, operations on the newly opened Baltimore & Ohio Railroad had been stopped but a single day. Nor could the canals ever successfully compete in the building of short lines or in furnishing service directly to the factory or mill of the shipper.

The railroad offered shippers tremendous advantages in speed. In 1852 rail shipments from Cincinnati to New York City took from six to eight days, about a third of the time required for service via the canals, Lake Erie, and the Hudson River. These several advantages resulted in heavy rail traffic of such freight as livestock, packing-house products, and general merchandise by the early fifties. The Erie Railroad was shipping huge quantities of milk, dairy products, and berries from the lower tier of counties in New York to New York City long before the completion of the line in 1851.

In the generation between the Treaty of Ghent and 1850 Americans had solved the problem of moving people and goods over the growing distances of their nation both cheaply and quickly in a variety of ways. Turnpike, wagon, and Concord stage had given way to canal, lineboat, and packet, to be supplemented, in turn, by stern-wheeler and side-wheeler on almost every river in the country. Before the typical turnpike had started to return its original investment—almost before the memory of the toasts and illuminated banners used to celebrate the opening of the Erie Canal had faded and long before the first federal safety regulation for river steamboats (1838)—another challenger appeared in the form of the flanged wheel, iron rail, and puffing locomotive. Perhaps the really significant theme in the early transportation history of this country was the tendency of each new type or phase of transportation to be so quickly challenged and largely supplanted by the next succeeding type or phase that no single form of transportation had a decent opportunity to grow. Fortunately for the railroad, this trend was halted by the middle of the nineteenth century. By then the railroad had achieved a position of dominance that it was to retain until early in the twentieth century.

The dream of Oliver Evans concerning practical rail transportation had clearly become a fact by 1850. Of course many improvements were still unknown to the American railways then; steel rail, extensive federal land grants, the automatic air brake, the adoption of standard-gauge track and standard time were all to be developed in future decades. Nevertheless, the American railroad of 1850 already had twenty years of growth behind it and was sure of its future in an expanding American economy.

3

Early Maturity: Expansion and War

The decade just before the Civil War was one of the most dynamic periods in the history of American railroads. In 1850 a broken skein of short lines stretched from Maine to Georgia, and a few stray strings of rail connected the Great Lakes and the Ohio River. In the fifties the discovery of gold in California, the lure of the trans-Pacific trade, and the new land-grant policy of the national government combined to push rail construction at a rapid rate. As a result a shift soon took place in the prevailing traffic patterns from north-south (following the Ohio and Mississippi Rivers) to east-west (following newly built rail lines). The railroads shared fully in the optimism, expansion, and prosperity so typical of the decade. By 1860 a railway network of over 30,000 miles served all the states east of the Mississippi quite adequately, and few locations of substantial population in the eastern third of the nation were far removed from the sound of the locomotive whistle.

The decade following 1850 was a time of many railroad "firsts." New York City, Philadelphia, and Baltimore all achieved their first rail connections with the west as the Erie, Pennsylvania, and Baltimore & Ohio railroads, respectively, reached Dunkirk, Pittsburgh, and Wheeling in 1851 and 1852. The first telegraphic control of

35

trains and the first locomotive used west of the Mississippi came in the same years. In 1856 the first railroad was opened in California, from Sacramento to Folsom; the first railroad bridge spanned the Mississippi; and the first railroad over seven hundred miles in length, the Illinois Central, was completed. And in 1860, Chicago, now served by eleven different railroads, had become America's leading, if not first, railroad center.

Railroads were the major "Big Business" on the American scene in mid-century. As the rail network grew from 9,000 to more than 30,000 miles in the decade, the total investment in the industry also more than tripled—from $300,000,000 to $1,150,000,000 in 1860. Few other institutions in the country did business on so vast a scale or financed themselves in such a variety of ways. Few other concerns in the nation employed such numbers of men so varied in skill. The energy of the railroad-builders was such that by the middle fifties the United States, with no more than 5 per cent of the total world population, had nearly as much rail mileage as the rest of the world. But the rate of rail construction in the pre–Civil War decade was quite uneven. In New England, where the rail promoters had almost overextended themselves in the forties, new construction was the slowest. To the west and south the amount and rate of construction increased. In New England the mileage increased in the decade by not quite 50 per cent, in the Middle Atlantic states it doubled, south of the Potomac and Ohio rivers the rail network more than quadrupled in mileage, and in the trunk-line region of the Old Northwest the increase was roughly eightfold. The most intense construction was centered in the three states of Ohio, Indiana, and Illinois. As much new rail was laid in Illinois in a single year, 1856, as had existed in the five states of the Old Northwest in 1850.

In New England well over half the new construction of the fifties was located in the three northern states of Maine, New Hampshire, and Vermont. The major new line completed through the three states in the decade was the Atlantic and St. Lawrence Railroad, a line built to connect Portland, Maine, with Montreal, 300 miles to the northwest. The man who dreamed of the railroad—and later built it—was John Alfred Poor (1808–71) a 250-pound Yankee from Maine. Poor's younger brother, Henry Varnum Poor (1812–1905), was the major spokesman for American railroads in the nineteenth century, first as long-time editor of the *American Railroad Journal* and later as publisher of the *Manual of the Railroads of the United States*. The Atlantic and St. Lawrence became a reality only after John Poor had made a dramatic dash from Portland to Montreal through a raging blizzard to convince the Montreal Board of Trade that his route was preferable to one terminating in Boston. The 292-mile line was completed in 1853 and was built, not to the standard gauge common in New England, but to the broad Canadian gauge of 5 feet 6 inches. Within a short time the road was leased to the Grand Trunk Railway of Canada, a line which by 1861 had reached the borders of Michigan via the shores of Lake Ontario and Toronto.

Farther south, four east-west main trunk lines were completed early in the fifties, two ending at Lake Erie and two reaching the Ohio River. The oldest of the routes to Lake Erie was the string of ten short lines paralleling the Erie Canal. These roads had been giving through service, of a sort, between Albany and Buffalo since the early forties. By 1853 they boasted a combined total of 542 miles of line, 150 woodburning locomotives, 1,700 freight cars, 187 first-class passenger coaches, and an advertised fourteen-hour passenger-train schedule between Albany and Buffalo that was rarely adhered to. The consolidation of the ten lines was completed in the summer of 1853, with the new company, the New York Central Railroad, boasting a sensational capitalization of $23,000,000. Erastus Corning, iron manufacturer of Albany, was the new president. As executive head of the Utica and Schenectady for twenty years, he had never taken a cent in salary, asking only for the privilege of supplying the railroad with all its needs in iron, steel, and rails, and the same profitable arrangement remained in effect with the New York Central until 1856, at which time a committee of stockholders decided that their company was too big for that sort of thing. Corning continued as president of the prosperous road until 1864, by which time the New York Central had caught the eye of its future leading man, Commodore Cornelius Vanderbilt (1794–1877).

In 1851, as the presidents of the ten little railroads were only beginning to think of consolidation, a second through line to the west, the Erie Railroad, was completed. Its president, Benjamin Loder, planned a record-breaking celebration to mark the completion of his line, at that time the longest railroad in the country. The two special trains which made the trip from Piermont to Dunkirk on May 14 and 15, 1851, carried 298 invited and assorted statesmen, including President Millard Fillmore, (1800–74), four members of his cabinet, and a number of presidential hopefuls, such as former Governor William H. Seward (1801–72), Stephen A. Douglas (1813–61), and Secretary of State Daniel Webster (1782–1852). At his own request the Secretary rode in a rocking chair securely fastened on an open flat car. Bundled up in a steamer rug, with a bottle of Medford rum for company, Webster said he wanted to miss none of the scenery. The route-long celebration included a number of speeches by Daniel, a parade (with nine thousand militia in dress attire), a seven-hour banquet at the overnight stop at Elmira, and a final feast at Dunkirk, where a three-hundred-foot table served up a variety of food and drink to the now groggy multitude. The new line had several years of fairly sober management before the shadow of Daniel Drew (1797–1879) fell across its path. In the early fifties Charles Minot (1810–66) was its able superintendent. Taking advantage of a parallel telegraph line set up earlier by Ezra Cornell (1807–74), Minot, on

September 22, 1851, used the telegraph for the first time to dispatch trains, a practice soon to be adopted by all the railroads.

Two other lines completed their original routes to the west in reaching the Ohio River late in 1852. In his first year as president of the Pennsylvania Railroad, John Edgar Thomson saw the completion of the road from Philadelphia to Pittsburgh on December 10, 1852. For two years the railroad used the inclined planes of the Allegheny Portage Railroad, but by 1855 a tunnel and Horseshoe Curve near Altoona made possible the exclusive use of locomotive-drawn trains. Farther down the Ohio the route from Baltimore to Wheeling was completed with the driving of the last spike on the Baltimore & Ohio Railroad on Christmas Eve, 1852. The first train entered Wheeling on January 1, 1853, initiating sixteen-hour passenger service from Baltimore, a trip that had earlier taken several days by turnpike stagecoach. Thus in the first years of the decade four rival western lines had completed their original planned construction. The race between Baltimore, Philadelphia, New York City, and Boston to tap western trade via the railroad had finally ended in a virtual tie. More important was the fact that the four trans-Allegheny railroads had united the states of the Northeast and the Northwest well before the crisis of secession.

Railroad construction in the Old Northwest in the fifties proceeded at a hectic rate, especially in the trunk-line region west of Pennsylvania. By 1860, Ohio was first in rail mileage in the nation, Illinois was second, and Indiana a strong fifth after New York and Pennsylvania. Much of the new construction in the region consisted of continuations of the trans-Allegheny lines of the East. Having extended their railroad to the Ohio at Wheeling, the owners of the Baltimore & Ohio sought the building of direct connecting lines westward to Cincinnati and St. Louis. Thomas Swann (1806–83) gave up the presidency of the B.&O. in 1853 but was soon president of the new Northwestern Railroad of Virginia, a line from Grafton to Parkersburg on a more direct route toward Cincinnati. West of Parkersburg the Marietta and Cincinnati Railroad was constructed, and west of Cincinnati the broad-gauge (six feet) Ohio and Mississippi Railway was built to complete the route to St. Louis. All three roads were finished by 1857, and the opening of the "American Central Line" from Baltimore to St. Louis was marked with several special trains carrying more than 2,500 persons. The entire route ultimately came under the control of the B.&O. Farther north the Pennsylvania Railroad had in 1856 acquired working control of a 468-mile route from Pittsburgh to Chicago, the recently completed Pittsburgh, Fort Wayne, and Chicago Railroad.

In Michigan another westward-looking line was completed as a trio of Yankees, John Murray Forbes (1813–98), James F. Joy (1810–96), and John W. Brooks (1819–81), completed the Michigan Central from Detroit to Chicago in May 1852. The same trio had just won a long-running dispute with some local Michigan farmers. The difficulty started when the railroad refused to pay more than half the value of killed livestock after it had gone to the trouble of fencing all its track. The farmers and sympathetic townspeople of Jackson County retaliated by filling journal boxes with sand, tampering with switches, greasing rails, and even subjecting cars and engines to fusillades of rocks and buckshot. After the burning of a depot the farmers were hailed into court on charges of conspiracy. A round dozen received jail sentences, despite the best defense efforts of New York lawyer William H. Seward. A second Michigan railroad, the Michigan Southern, also reached Chicago early in 1852. This line was to become part of the Lake Shore & Michigan Southern, a road which Commodore Vanderbilt was to use as the western extension of his New York Central.

Chicago, a city of thirty thousand residents by 1850, was the focal point of much of the railroad-building in the decade. William Butler Ogden (1805–77), wealthy realtor and first mayor of the city, advocated in the middle forties that railroads be built to replace the growing plank-road network that served the city. Few Chicago merchants agreed with him, but by 1848 he had opened the first section of the Galena and Chicago Union Railroad (later a part of the Chicago & North Western), which was intended to serve the lead-mining region west of Chicago. In 1850 the 43-mile line was grossing $48,000 a year, had an operating ratio of less than 50 per cent, and boasted a new depot with a tower from which Ogden, using a marine telescope, spotted incoming trains for the benefit of passengers on the platform below. Within a few years most of the first-class hotels in the city had comparable cupolas in which men spied out incoming trains and boats so that omnibuses could be sent to meet the arriving passengers. Since each of the new rail lines had its own individual depot, the need for a transfer service between stations was soon obvious. In 1853 Frank Parmelee (1816–1904) established a transfer service long noted for its handsome rigs, excellent horseflesh, and genial and distinctively dressed drivers.

In the fifties many radiating lines were built out of Chicago. The Forbes group pieced together the Chicago, Burlington & Quincy to reach two points on the Mississippi, the St. Louis, Alton and Chicago was built to St. Louis, and the 180-mile Chicago, Rock Island & Pacific was finished to Rock Island in the record time of under two years by Henry Farnam (1803–83) for the company's president, John B. Jervis, builder of the Mohawk and Hudson. The Rock Island is remembered for building the first railroad bridge across the Mississippi (1856), a bridge both legally and physically threatened by rival steamboat interests. Abraham Lincoln (1809–65) was among the lawyers who won the subsequent lawsuit for the railroad.

Of the eleven railroads that served Chicago in 1860 the Illinois

Central was the longest and perhaps the most important. Unlike most railroads in Illinois, it was a north-south line running the length of the state. Much like the shape of a thin wishbone, one section of the road ran south from Chicago and a second extended south from Dunleith, the two lines meeting at Centralia to continue down to Cairo. In 1850 Senators Stephen A. Douglas of Illinois and William R. King (1786–1853) of Alabama maneuvered through Congress the first land-grant act aiding the building of a railroad. The law granted six alternate sections of land per mile of railroad to the Illinois Central in Illinois and the connecting Mobile and Ohio in the states of Mississippi and Alabama. In answering critics of the proposed land grant, Senator King, later Vice-President under Pierce, said: "This is an immense grant . . . but it will be there for five hundred years; and unless some mode of the kind proposed be adopted, it will never command ten cents." The Illinois Central was soon profitably selling thousands of acres of its land and bringing hundreds of new settlers into the state. When the Illinois Central completed its main line in 1856, its seven hundred miles of track made it the longest railroad in the world.

South of the Ohio and Potomac rivers several important railroads were completed in the decade before the Civil War. The sister line of the Illinois Central in the South, the Mobile & Ohio, finished its 483 miles of track from Mobile to Columbus, Kentucky, in the spring of 1861, just ten days before the start of the Civil War. Under the leadership of James Robb (1814–81), with some assistance from Judah P. Benjamin (1811–84), a second line to the North, the New Orleans, Jackson and Great Northern, was built from New Orleans to Canton, Mississippi. A connecting road, the Mississippi Central, continued the route into Kentucky. In the same years the uncouth but highly regarded James Guthrie (1792–1869) of Louisville, after retiring from Pierce's cabinet in 1857, completed the Louisville & Nashville Railroad. The major east-west route built in the decade consisted of four lines: the

Memphis and Charleston, running from Memphis to Chattanooga; the East Tennessee and Georgia, from Chattanooga to
Knoxville; the East Tennessee and Virginia, from Knoxville to
Bristol; and the Virginia and Tennessee, which ran to Lynchburg,
Virginia. Every southern state added substantial railroad mileage
in the decade. Virginia, because of the extension of new roads to
the west (the B.&O. and the road to Bristol), easily led the South
in rail construction and ended the decade with well over 1,700 miles
of line, thereby demoting Georgia to second place for a few years.

The American rail network of 1860 could be divided, on the
basis of mileage, into nearly equal thirds: ten thousand miles in
New England and the Middle Atlantic states, eleven thousand
miles in the Middle West, and nine thousand miles in the South.
But in average length, cost of construction, and volume of traffic
the four hundred railroad lines in the three regions varied widely.
In the Northeast, where traffic was heavy and the distances com-
paratively short, the nearly two hundred different lines averaged
about fifty miles in length and were valued (in 1860) at an average
of $48,000 per mile. In the Middle West the longer distances and
often lighter construction and traffic produced figures of just over
a hundred miles for average length of each line and a cost of
$37,000 per mile. Southern lines averaged perhaps ninety miles in
length and because of light construction, easy terrain, and low
traffic density were valued at only $28,000 per mile on the eve of
the Civil War.

Technical advances in equipment and operation came naturally
in the decade. Eleven different gauges were in use in the North,
but by 1860 the standard gauge, 4 feet, 8½ inches, was by far the
most common. The great bulk of the southern railroads were still
of the 5-foot gauge. In 1860, T-rails were the general rule, but a
few branch lines and some roads in the South still had iron-capped
wooden rails. Steel rails came into use only after the war. Most
railway superintendents in the decade were ordering American-

type (a swiveled four-wheeled truck in front plus four drivers) locomotives with functional cowcatcher, large headlight, and balloon stack. This type of engine (4-4-0) in 1860 had a name rather than a number, cost $8,000 to $10,000 to build, used wood or possibly coal for fuel, cost well under a dollar a mile for total operating expense, and was the pride and joy of the engine crew assigned to it. For heavy freight duty Ross Winan's camel-back was still common on the B.&O., and a new Mogul-type (2-6-0) freight engine appeared in 1863.

The best coaches now came equipped with corner toilet, water tank, and newsboy. During the last years of the decade at least four men—Webster Wagner (1817–82) for the New York Central, Edward C. Knight (1813–92) for the B.&O., T. T. Woodruff (1811–92) for the Terre Haute and Alton, and George M. Pullman (1831–97) for the Chicago and Alton—were experimenting with sleeping cars. Pullman had earlier made a name for himself in Chicago by successfully lifting the four-story all-brick Tremont House several feet to the new street level with the aid of 1,200 men and some 5,000 jackscrews. Pullman's sleeping car was to become the accepted model because he had, with equal ingenuity, worked out an upper berth which would disappear during the day while holding the bedding for both upper and lower beds. Night travel had become the accepted thing by 1860, but Sunday trains were still matters of dispute, especially in the southern states, until after the Civil War. Dining-car service was not yet available, the traveler having to be satisfied with hurried snacks bought at railroad eating houses during short scheduled stops.

The fifties brought both a great expansion in railroad mileage and an increase in the number of railway accidents. Faulty track maintenance, poorly built bridges, careless railroad employees, and often the total absence of the most elementary safety precautions all contributed to the high catastrophe and casualty rate. As railroad superintendents and directors urged on the hectic rate of railroad-building, their motto often seemed to be: "Let's lay more track and to hell with maintenance." Carelessness by the engineer clearly caused the 1853 wreck at Norwalk, Connecticut, when a train ran on to an open drawbridge, killing 45 persons. The absence of any traffic control over the crossing of two rail lines in Illinois caused another wreck later that same spring, bringing death to 18 and injury to nearly 60. The total for the year was more than a hundred major rail accidents, with 234 passengers killed and 496 seriously injured.

American railroad bridge-builders were among the cleverest in the world. John A. Roebling's (1806–69) double-deck suspension bridge near Niagara Falls, which he had begun by flying a cord-carrying kite across the chasm, was finished in March, 1855. But most of the wooden "Howe truss" bridges built in the fifties were not as safe as the Niagara bridge, the Erie Starucca Viaduct, or the enduring stone Thomas Viaduct on the B.&O. Too often the wooden bridges of the day grew tired, the trains fell through, and the car stoves ignited the resulting debris.

Nor was the American rail system on the eve of the Civil War really an integrated network. Gauge diversity was one of the most serious handicaps to through service. In 1861, because of different gauges, eight changes of cars were necessary for a trip from Charleston to Philadelphia. Impediments to through traffic were caused not only by the absence of over-all planning but also by the presence of strong local economic interests. Tavern keepers, teamsters, and porters were happy that not a single rail line entering either Richmond or Philadelphia made a direct physical connection with any other railroad entering the city. Gaps in service were also to be found at the major rivers. While the Mississippi had been bridged before the war, neither the Ohio nor the Potomac was crossed by a railroad bridge in 1861. A major result of this variety of gauge, lack of transfer facility, and absence of important bridges was that in the railroad service of a century ago nearly ev-

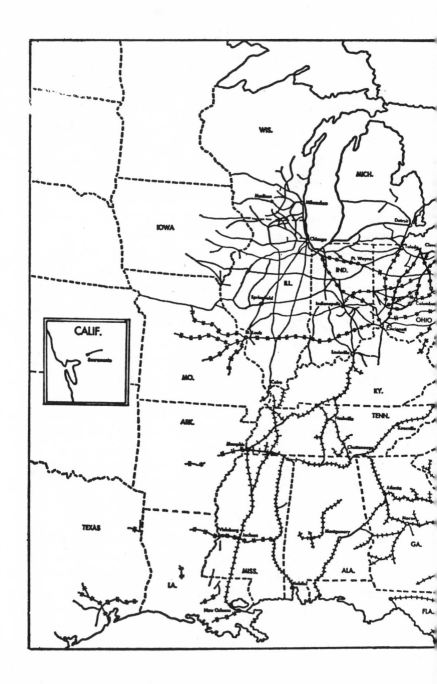

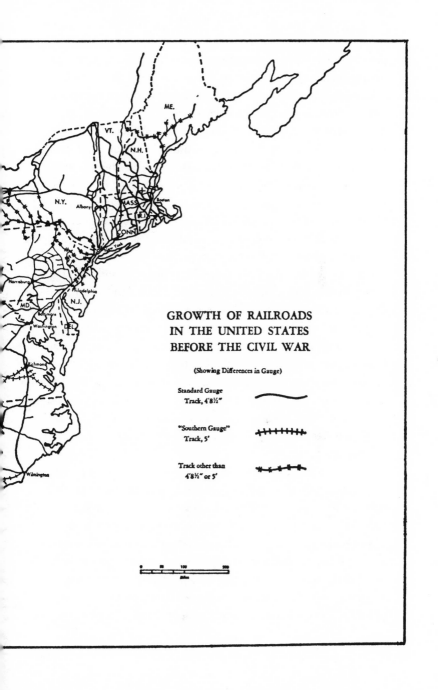

GROWTH OF RAILROADS
IN THE UNITED STATES
BEFORE THE CIVIL WAR

(Showing Differences in Gauge)

Standard Gauge
Track, 4'8½"

"Southern Gauge"
Track, 5'

Track other than
4'8½" or 5'

ery freight car belonged to the road on which it was running. True physical integration of the railroad network would have to wait until well after the Civil War.

Imperfect as the network was, rail transportation in the period from mid-century to 1865 provided the nation with vast improvements in the movement of both goods and people. Railroad freight rates dropped substantially during the fifties. While the New York Central, Erie, and Pennsylvania lines had averaged from 2.4 cents to 3.5 cents a ton-mile in 1853, by 1860 the average rate for the three roads was less than 2 cents. As the rail network grew rapidly in the region north of the Ohio River, much of the western domestic commerce which had been accustomed to a southern water movement increasingly shifted to eastern markets that were reached by rail and the Great Lakes. Between 1852 and 1856 the arrival of wheat in Chicago increased ninefold, and comparable shipments of corn quadrupled. This grain came to Chicago via the new railroads, and much of it went on to the East the same way. New Orleans and the river boats even lost some of their cotton trade to the railroads. When the Western and Atlantic Railroad and the Memphis and Charleston were completed in the fifties, much of the cotton that once had descended the Tennessee and Cumberland rivers to an eventual market in the Crescent City went instead by rail to increase the exports of Savannah or Charleston.

By 1860 the canal packets and river steamers had lost much of their passenger traffic, and many railroads had improved and sped up their schedules in the fifties. On the eve of the Civil War the railroad passenger could travel from St. Louis to Boston in forty-eight hours or from New York to Charleston in sixty-two hours. He could buy a coupon ticket good for the entire trip from Bangor, Maine, to New Orleans, check his baggage the entire distance, and note the performance of his train and train crew in the railroad schedules and guidebooks recently made available.

Despite its limitations the American rail network had by the Civil War furnished the nation with a transportation system far beyond anything made available by earlier forms of internal improvement. Turnpikes had never been built in any significant numbers in either the southern or western states, being limited chiefly to New England and the Middle Atlantic region. Canal-building had been somewhat more extensive in many states, but in 1850, when the mileage was near its peak, only fourteen states could claim as much as twenty-five miles of canal. By contrast, in 1860 every state save Minnesota and Oregon had railroad mileage, and twenty-nine of the thirty-three states had more than a hundred miles of line. More than a dozen lines were reaching west from the Mississippi into the states of Iowa, Missouri, Arkansas, and Louisiana. In fact, on the eve of the Civil War the western fingers of the rail network were very close to the frontier line. With the exception of small corners of Minnesota, Iowa, and Kansas, the Ozarks of Missouri and Arkansas, and much of eastern Texas, the western population of the Mississippi Valley was at least partially served by the railroad.

In the fifties many Americans were also dreaming of and making plans for a railroad to the Pacific. Even before the Mexican War the arguments of Asa Whitney (1797–1872), New York merchant and world traveler, for a railroad to the Pacific via the South Pass had gained modest support in Congress. The acquisition of California, the discovery of gold, and the prosperity of the period greatly increased the interest in such a project. Milwaukee, Chicago, St. Louis, Memphis, Vicksburg, and New Orleans all aspired to become the eastern terminus of the proposed rail route. In March, 1853, Congress appropriated $150,000 for the survey of four possible routes to the coast. Between 1853 and 1855, three central and southern routes were examined by U.S. Army engineers, and a northern route was surveyed by Isaac Stevens (1818–62), governor of the Washington Territory. The Gadsden Purchase of 1853, which had been urged upon President Franklin Pierce (1804–69) by Secretary of War Jefferson Davis (1808–89), helped southerners feel that their favorite route would be selected. Early in 1854 Senator Douglas of Illinois, who clearly preferred a central route starting from Chicago, countered with his Kansas-Nebraska Bill. Douglas hoped that the organization of territories west of the Missouri River would improve the chances of having the central rail route selected. In the resulting flare-up of antislavery and proslavery passions the Pacific railroad was largely forgotten. The Panic of 1857 definitely ended any thought of such a railroad in the fifties. The depression of the late fifties checked rail construction moderately in the rest of the nation. As the rate of expansion declined, attention was frequently given to consolidation and improvement of rail management. On the eve of the Civil War American railroads were ready for, were waiting for, an increase in business. The war brought that increase.

If the increase in business which came with the Civil War improved the economic health of the American railroads, the railroads, in turn, materially aided the Union cause. Had the war started a decade earlier, before the substantial rail construction of the fifties, both the military and economic contributions to northern victory made by the bottled-up Ohio and upper Mississippi valleys would have been significantly smaller. In the quarter of a century before 1850 the competition for the commercial allegiance of the Northwest had resulted in a near tie between the Erie Canal boats from the Northeast and the river steamers from the South. In the fifties the economic victory had gone to the Northeast as hundreds of miles of new railroad iron crossed the granite barrier of the Alleghenies to enter the Old Northwest. Iron rails and puffing locomotives helped the Northwest decide that it would support the Union.

As the Union and the Confederacy faced the prospect of war in the spring of 1861, their respective rail systems offered more in the way of contrast than comparison. With roughly nine thousand miles of line, the eleven Confederate States could claim a small third of the nation's rail mileage. Their roads were more lightly constructed, carried a correspondingly smaller volume of traffic, and employed no more than a fifth of the country's railroad employees. A traditional dislike for mechanical pursuits in the South had resulted in the employment of many northerners on southern roads. With the outbreak of war, a number of these northerners returned home, and many of the remaining workers were viewed with suspicion. There also existed a great disparity between northern and southern rolling stock and motive power. The Pennsylvania and Erie railroads had nearly as many locomotives as the entire Confederacy. In 1861 these two lines plus the New York Central and the Baltimore & Ohio possessed roughly as much motive power and rolling stock as could be found in all the states south of the Ohio.

In the wartime procurement and maintenance of rolling stock, motive power, and track the South was at a distinct disadvantage. Southern railroad officials had always bought most of their rail equipment in the North. Virginia, South Carolina, and Tennessee produced a fair number of railroad cars, but the productive capacity of Pennsylvania was probably twice that of the entire South. The states north of the Potomac had a dozen locomotive plants for every one in the South. The few engine-building facilities that did exist in the Confederacy, such as the Tredegar Iron Works at Richmond, were soon pressed by the Davis government into concentrating on the production of ordnance. As early as the spring of 1862 southern railroad officials were predicting the complete breakdown of their service because of shortages of rolling stock and motive power.

The problem of track maintenance was acute. The South produced some rails, but its production of twenty-six thousand tons in 1860 was only a ninth that of the North. Southern rail presidents had long preferred English to northern rails, claiming them to be both superior and cheaper. As the northern blockade stopped this source and as ordnance production slowed Confederate rail production, southern railroad men began to hoard their iron. Soon they were robbing branch lines to maintain their main stems. Before the war was over the Confederate government was doing the same thing with whole lines, seizing railroads in Texas, Florida, and Georgia. Had the government of Jefferson Davis earlier faced up to its transportation problem and established as vigorous a policy of railroad control as it did in the field of ordnance, southern railroads might well have provided a more adequate support for the Confederate military effort.

With the exception of some minor railroad activity in the Mexican War and the Crimean War, the Civil War was the first conflict in which railroads played an important role. As the war started, the *American Railroad Journal* predicted that the majority of the railroads would be unaffected by the conflict. The summer weeks of 1861 proved this prediction to be the understatement of the year as the logistical and material needs of both armies became apparent. Southern lines leading toward the Virginia and Tennessee fronts were soon overwhelmed with a flood of excited soldiers and ancient ordnance moving northward. The Petersburg Railroad south of Richmond had a record-breaking operating ratio of less than 28 per cent in 1862, and the Wilmington and Weldon was able to pay a 31 per cent dividend in 1863. In neutral Kentucky during the first days of the war, the Louisville & Nashville Railroad had such a heavy traffic in Confederate purchases from the North that the railway's president, James Guthrie, finally imposed a ten-day embargo to permit the road to clear its lines. Much of the southern rail prosperity was unreal because of serious inflation in the Confederacy. In the South, railroad purchasing agents went crazy as the prices for their needs went up. Between 1861 and 1864 lubricating oil rose from $1.00 per gallon to $50.00 per gallon, car

A wrecked locomotive at Richmond, Virginia, in April, 1865, was typical of the ruined railroads of the South. The Confederate capitol can be seen at the right of the smokestack. (Courtesy, Library of Congress.)

Rails stored at Alexandria, Virginia, for the U.S. Military Railroads. By the last months of the Civil War almost any railroad president in the Confederacy would have pledged his railroad for the hundreds of tons of good railroad iron shown here. (Courtesy, The National Archives.)

Hanover Junction, Pennsylvania, in 1864, on the route to Gettysburg. The crowd may be explained by the presence of the tall gentleman just to the right of the cowcatcher, who some believe to have been President Lincoln. (Courtesy, Association of American Railroads.)

"The 9:45 Accommodation, Stratford, Connecticut." This 1867 painting by E. L. Henry clearly indicates the place of the depot in American life a century ago. (Courtesy, Metropolitan Museum of Art, Bequest of Moses Tanenbaum, 1937.)

wheels from $15.00 to $500.00 each, mechanics' wages from $2.50 to $20.00 a day, and coal from 12 cents to $2.00 a bushel. Railroad rates could not be pushed up as fast as these mounting costs. In 1864, John P. King (1799–1888), president of the Georgia Railroad, wrote of his line: "The more business it does, the more money it loses, and the greatest favor that could be conferred upon it—if public wants permitted—would be the privilege of quitting business until the end of the war!"

In the North, rail prosperity was more substantial. The Illinois Central throughout the war carried heavy trainloads of recruits, horses, forage, ordnance, and wounded and returning veterans up and down its line. Even though his line, as a land-grant road, carried this huge government business at greatly reduced rates, President William H. Osborn (1820–94) was able to announce dividends of 8 per cent in 1863 and 1864 and 10 per cent in 1865. The Erie finally paid a dividend in 1863, and the New York Central raised its dividend rate from 6 per cent to 9 per cent between 1860 and 1864. As the war progressed, these two lines saw their combined freight tonnage increase from a figure slightly less than that of the Erie Canal in 1861 to a figure 40 per cent larger than canal tonnage in 1865. Most other northern lines had tonnage increases and dividend rates that were equally good. Forgetting its earlier prediction of the effect of war on the country's railroads, the *American Railroad Journal* in January, 1864, called the year 1863 "the most prosperous ever known to American railroads."

In addition to the important day-to-day operations of moving military personnel and matériel, the railroads, both Union and Confederate, were frequently used for rapid and massive troop movements. As General George McClellan's (1826–85) army slowly approached Richmond in early June, 1862, Jefferson Davis and Robert E. Lee (1807–70) felt secure enough to send two brigades (10,000 men), under Generals William H. C. Whiting (1824–65) and Alexander Lawton (1818–96), to strengthen

Thomas "Stonewall" Jackson (1824–63) in the Shenandoah Valley. The entire movement was by rail. The same summer, in late July 1862, General Braxton Bragg (1817–76) moved his entire army of roughly 30,000 men from Tupelo, Mississippi, to Chattanooga. The total movement took little more than a week, although the indirect route, via Mobile, Montgomery, and Atlanta, involved a journey of 776 miles over six different railroads. A year later Bragg was reinforced during the Battle of Chickamauga as General James Longstreet's (1821–1904) entire First Corps of the Army of Northern Virginia (between 12,000 and 15,000 men) was moved by rail over ten railroads and 900 miles from Richmond to northern Georgia. The railroads of the South were still in fair shape in 1863.

The greatest single movement of Union troops came in late September and early October, 1863, as Lincoln's government sought to reinforce General William S. Rosecrans (1819–98), now besieged in Chattanooga after his defeat at Chickamauga. Secretary of War Edwin M. Stanton (1814–69) proposed to Lincoln's inner cabinet that the War Department send by rail 30,000 men from the Army of the Potomac to break the siege of Chattanooga. The Secretary estimated that the job could be done in five days. Lincoln was so skeptical that he said to Stanton: "I will bet that if the order is given tonight the troops could not be got to Washington in five days." Stanton was nearer right than the President. In eleven and one-half days 25,000 men and 10 batteries of artillery, with all their horses, traveled 1,200 miles from Washington to the banks of the upper Tennessee. The route of the 30 trains of nearly 600 cars was via Harper's Ferry, Columbus, Indianapolis, Louisville, and Nashville.

Another outstanding rail contribution to Union victory was made as the forces of General William T. Sherman (1820–91) left Chattanooga for Atlanta in the spring of 1864. When Sherman reached Atlanta, his supply line consisted of a series of single-track railroads stretching back 473 miles, via Chattanooga and Nashville, to Louisville. For months the normal daily flow of supplies sent southward to the 100,000 men and 35,000 animals of Sherman's army consisted of 16 trains of 10 cars (ten-ton capacity) each, or a total of 1,600 tons. With such a slender supply stem, Sherman had good reason to worry about such southern raiders as Nathan Bedford Forrest (1821–77).

During the Civil War neither government, except in particular emergencies, ever took complete control of its railroad establishments. The fact that the Confederacy never exercised an effective supervision over its inferior railroad system clearly contributed to the failure of the war effort of the South. In the North, where the basic vitality of the railway network made strong regulation less necessary, Lincoln's government was more willing to act. Legislation of January 31, 1862, permitted the government, if necessary, to seize and operate northern railroads and also to control the operation of captured southern lines. Daniel C. McCallum (1815–78), strict disciplinarian, former superintendent of the Erie, and perhaps the only general in the Union Army who was a popular poet, was placed in charge of the government's railroad operations. The General never used his sweeping powers fully, functioning instead as a sort of liaison officer between the government and the northern railroad presidents. McCallum was also director of the United States Military Railroads, which grew from a seven-mile line in 1862 to a system of 2,105 miles in line, 419 engines, and 6,330 cars at the end of the war. Most of this network supported the western operations of Sherman, and much of the mileage was composed of former Confederate railroads.

In the eastern theater of operations construction activities, especially the bridge-building magic of Herman Haupt (1817–1905), were of critical importance. A West Point graduate, Haupt had early left the army for a variety of railroad experience in the years before the Civil War. Unlike McCallum, whose admirable administration of railroads was composed of paper work, clean-cut decisions, and formal instructions, General Haupt was a leader of men willing to get his shirt sweaty and his boots muddy as he built bridges. His bridges were built in record time to carry heavy loads, even though, as President Lincoln said, they seemed to be constructed of nothing more than "beanpoles and cornstalks."

As the rival armies struggled to possess such vital railroad junctions as Corinth, Chattanooga, and Atlanta, a great deal of railroad destruction occurred. With most of the fighting in the southern states, northern railroads, with the exception of some exposed sections of the Baltimore & Ohio and the Hannibal & St. Joseph, generally were not molested. All the Confederate states suffered damage, but the havoc was greatest between Virginia and Mississippi. One of the earliest—and certainly the most publicized—railroad raids of the Civil War was the "Andrews Raid" of April 1862, in which a group of disguised Union soldiers attempted to wreck the Western and Atlantic Railroad after stealing the locomotive "General." The raid was colorful but totally ineffective.

The high point of rail destruction came during the five-month scourge which Sherman's troops gave Georgia and the Carolinas in 1864–65. Confederate raiders never acquired the pure destructive skill of the more mechanically minded northern soldiers. In the summer of 1865 a former Confederate captain told Sidney Andrews (1835–80) this about rail destruction: "We could do something in that line, we thought, but we were ashamed of ourselves when we saw how your men could do it." Southerners also contributed to their own railroad ruin as they occasionally tore up their own track before retreating, and Confederate soldiers were frequently guilty of rough usage of southern rail equipment. As Longstreet's troops rode the rails in September 1863, on their way to help out at Chickamauga, the pent-up men pried off the side walls of their boxcars to improve both view and ventilation.

The various destructive forces left southern railroads a shambles by the end of the war. The conflict had completely destroyed or crippled well over half the Confederate railways. Twisted rails, burned ties, destroyed bridges, gutted depots and shops, and lost or dilapidated rolling stock were the normal heritage of war for most southern lines. When Chief Justice Salmon P. Chase (1808–73) visited North Carolina in May, 1865, he was provided with a train which the Washington correspondent Whitelaw Reid (1837–1912) described as "a wheezy little locomotive and an old mail agent's car, with all the windows smashed out and half the seats gone." Some of the lines were out of service a long time. Operation of the 104-mile rail link between Savannah and Charleston was not resumed until March, 1870. The rehabilitation program was helped by the federal government in both the early return of the southern lines, which General McCallum had renewed, and in the sale (on short-term credit) of government-owned rolling stock and equipment. By Christmas, 1865, most southern lines, ruined though they were, were offering some sort of service.

With the exception of the southern lines, American railroads were generally in excellent shape in 1865. War needs had caused the nation's network to expand to some 35,000 miles by the end of the conflict. Most of the 4,500 miles of track built since 1860 was located in either the Middle Atlantic states or the Old Northwest. At war's end the northern railroads could be proud of their role in the successful struggle to preserve the Union. Their rapid construction in the fifties had assured a united North when war came. In addition to the obvious contributions of a strategic and tactical nature, the railroads had also helped make possible the successful marketing of lush northern harvests to a hungry Europe ready to exchange credits and wartime neutrality for American grain.

The Civil War strengthened American railroads. The activities of the cohorts of McCallum and Haupt brought new efficiency in both track and bridge construction, and the bulk of the wartime traffic brought prosperity to most northern lines. Finally, the stimulation of wartime demands resulted in an enforced and healthy co-operation among railroads, created new methods of handling mail on trains, and forced the beginnings of a change from iron to steel rails and a comparable shift in fuel from wood to coal. American railroads achieved a measure of maturity in the expansion of the fifties and the challenges of war which followed. In 1865 they stood on the threshold of their golden age.

4

The Rails Move West

In the generation before the Civil War American railroads had built a network which served the eastern half of the nation fairly well. In the year of Appomattox the western fingers of this rail network reached nearly to the edge of the frontier in Wisconsin, Iowa, Missouri, Arkansas, and Texas. West of the frontier of the 1860s stretched the Great Plains, or the "Great American Desert," a region which Horace Greeley (1811–72), after his overland trip of 1859, described as destitute of human inhabitants because of the scarcity of wood and water. Greeley, Abraham Lincoln, and other commentators of the sixties believed that it might take a century to settle this last frontier. They were proved wrong in the 1870s and 1880s as the rapid transcontinental rail expansion of those years moved ahead of the frontier line and pulled millions of Americans into the western territory. Only a half-century was required to see the admission of the last of the forty-eight states in the main continental area.

During the age of the railroad (or since the 1820s) twenty-four states were admitted to the Union. Of these, half were admitted before 1865 and the remaining dozen after the Civil War. At the time of the admission of the first group of states (Arkansas, 1836, to Nevada, 1864), a total of only four hundred miles of railroad was in operation in the twelve states. This mileage was found only in Michigan and West Virginia, and railroads were in actual operation (at the date of admission) only in the latter state. West Virginia was really an exception since in 1863 she could scarcely be considered a frontier state.

After the Civil War the railroads quickly moved ahead of the frontier line. In the last dozen states joining the Union (Nebraska, 1867, to Arizona, 1912) the railroad clearly preceded both extensive settlement and admission. Except for Nebraska and Colorado, railroad service was available in each of the territories at least a decade before admission and in the case of the last five states admitted (Wyoming, Utah, Oklahoma, New Mexico, and Arizona) railroads preceded admission by more than twenty years. At the time of their respective admissions only one of the dozen states, Nebraska, had substantially under 1,000 miles of railroad, and the group as a whole could claim 23,000 miles of line as they joined the Union. In 1865 the Great Plains and Mountain West (west of and including the tier of states from North Dakota southward to Texas) had only 960 miles of line. In the next half-century this rail mileage was increased nearly a hundredfold to more than 90,000 miles. Clearly, the railroad played a significant role in the last American West.

As the Civil War veteran or the homesteader viewed the last frontier in 1865 he saw a vastly different West than that settled by the pioneer in the decades before the war. That earlier frontier east of the Mississippi had been one which, with all its hardships and trials, supplied the settler with plenty of water, timber, and land. This new land of prairie, stretching from the Missouri River to the mountains, was sparsely wooded, possessed few streams that were navigable, except in the wettest weather, and had a normal annual rainfall so low that it was the despair of settlers who had earlier farmed in the East. Short of water and timber, the last frontier on the Great Plains had only an abundance of land, warlike Indians, and distance. The windmill, new dry-farming methods, and, eventually, irrigation projects did much to overcome the problem of insufficient rainfall and seasons of drought. Barbed-wire fencing and the sod house helped the farm family to solve the problem of timber. The six-shooter, the United States Army, and the eventual destruction of the buffalo reduced the Indian menace. The western railroads solved the problem of distance, bringing many settlers to the new lands and opening up eastern markets for western agricultural production.

Many of the railroads built in the western half of the United States in the years after the Civil War were headed for the Pacific Coast or at least had high ambitions in that direction. In 1880, of the dozens of railroads in the states admitted to the Union after the Civil War, nearly a third had either the word "Pacific" or "Western" included in their official corporate name. There were few reasons for the western railroad built in the years just after Appomattox to pause on the Great Plains. There was little to stop for, except a thin trickle of westward-moving population and the beginnings of eastward-moving shipments of hides and cattle. The major objective was the Pacific or at least the mining towns and camps of the Rockies. As a result nearly all lines pushed westward as directly as the terrain would permit. A glance at the lines west of the ninety-eighth meridian on a railroad map of 1890 will show how few lines ran north or south. Since the average western railroad built after the Civil War found little immediate traffic, it was necessary for society to subsidize much of the first construction. Ambitious towns and cities, states, and the national government shared in granting these necessary subsidies.

The first major rail construction to the West after the Civil War was the completion of a transcontinental line. Dreamed of by Asa Whitney in the forties, the project was actively sponsored by rival groups in the prosperous fifties. But the intense sectional rivalry of the decade plus the Panic of 1857 kept the proposals in deadlock in the years before Fort Sumter. The Civil War caused a scarcity of capital, materials, and labor for western construction, but at last definite and specific plans could be agreed upon. True to the Republican platform of 1860, that a Pacific Railroad had to be built, Congress passed and Abraham Lincoln signed on July 1, 1862, the first Pacific Railway Bill.

The railroad legislation stated that the line to the Pacific was to be built by two companies: the Union Pacific Railroad Company to build westward from the Missouri River and the Central Pacific to build eastward from Sacramento, California. The Central Pacific had been organized and chartered in 1861. The new Union Pacific was to be capitalized at $100,000,000, and 162 commissioners were to solicit stock subscriptions. Both companies were to receive ten alternate sections of public lands (increased to twenty sections in 1864) for each mile of track. In addition, each of the two lines received a thirty-year government loan in United States bonds. The amount of the loan per mile of track varied with the terrain, $16,000 per mile being granted across the plains, $48,000 per mile for the high mountain area, and $32,000 per mile for the plateau region between the Rockies and the Sierras. Both the Union Pacific and the Central Pacific were permitted to issue first-mortgage bonds up to the amount of the government loan or subsidy.

By the fall of 1863 the required initial stock subscriptions in the Union Pacific had been made and the company was formally organized. General John A. Dix (1798–1879), a prominent citizen of New York and subscriber to fifty shares, was elected president. Dix never took an active interest in the management of the company, and this soon fell into the hands of its vice-president, Dr. Thomas C. Durant (1820–85), a railroad promoter who earlier (with Henry Farnam) had built the Chicago, Rock Island & Pacific. Dr. Durant was already much better known for his skill in stock manipulation than for his practice of medicine. Ground for the new line was first broken at Omaha, Nebraska, on December 2, 1863, with the merchant-author-eccentric George Francis Train (1829–1904) making the major speech. But labor was scarce, the connecting rail line east across Iowa was far from completion, and new stock subscriptions were slow in appearing. Months passed with little real progress. In the summer of 1864, however, Dr. Durant did succeed in getting Congress to amend and sweeten the original Pacific Railroad Act of 1862.

As plans for construction west of Omaha were made in the fall of 1864, Peter A. Dey (1825–1911), chief engineer for the line, estimated that the first hundred miles of road could be built for $30,000 a mile. Dr. Durant instructed Dey to raise his figures to permit lower grades and broader embankments. Following instructions, Dey changed the specifications and raised the estimate to $50,000 a mile. Durant proceeded to let the contract to Herbert M. Hoxie (1830–86), who really represented the Credit Mobilier. The contract was for $60,000 a mile but contained the cheaper specifications from Dey's original estimate. Dey, an honest engineer and a man of integrity, resigned. Early in 1866 the Union Pacific selected a new chief engineer, General Grenville M. Dodge (1831–1916). General Dodge was an equally capable engineer and was also more amenable to the ways of Dr. Durant and corrupt construction companies. It was Dodge who, while surveying Iowa railroads, had convinced Abraham Lincoln in 1859 that Council Bluffs, Iowa, would be the best eastern terminus for a railroad to the Pacific.

General Dodge built the Union Pacific. Corrupt or not, the Credit Mobilier (as a separate company composed of Union Pacific stockholders who paid themselves fancy prices for supplies and construction contracts) supplied the necessary incentive and helped provide the necessary funds. By mid-September 1866, some 180 miles of track had been laid, and by the end of the year the base construction camp was at North Platte, Nebraska, 293 miles west of Omaha. Dodge, in complete charge, directed the construction operation as if it were an army. It was in reality much like an army, both in size and personnel. At the climax of the race with the Central Pacific probably 10,000 workers, and nearly as many draft animals, were engaged in grading, building bridges, and laying track. The work gangs were composed of ex-soldiers from both armies, ex-convicts from the East, Irish from New York City, and a scattering of residents from the plains and mountains. Reluctant to see their buffalo country disturbed by iron rails, the Indians resisted the invasion across the prairies of Nebraska and on into the higher land of Wyoming. Most of the workers were forced to be as adept with revolver or carbine as with pick or shovel. The Army supplied only partial protection, for as General George Crook (1829–90) remarked, It was not easy for one soldier to surround three Indians.

In spite of Indians, the high cost of supplies, and the lack of water and timber, the road was constructed. General Dodge's assistants, the Casement brothers, Daniel and General Jack (1829–1909), pushed the work gangs at a rapid pace. In 1867–68, a total of 450 miles of track were laid, and by the early months of 1869 the advance grading crews of the Union Pacific and the Central Pacific, in their eagerness to build as much subsidized road as possible, had passed each other with parallel lines. The construction was often hurried with flimsy bridges, narrow embankments, and improperly ballasted track. General Dodge himself admitted that his company's greedy insistence on continued construction in the winter months often doubled or even tripled building costs. The haste in construction was also caused by a public that wished to see the job completed.

As the track was pushed ever farther west the temporary base camps of the railroad were immediately augmented by the flimsy shacks and tents of saloon keepers, prostitutes, gamblers, and sharpers of the roughest sort. When the winter or base camps moved farther west most of these "hell on wheels" communities left nothing but a littered void on the prairie. Some few of the camps, such as North Platte, Julesburg, Cheyenne, and Laramie, settled down for a slow growth toward respectability. All of them had originally been, as Robert Louis Stevenson described them, "roaring, impromptu cities full of gold and lust and death."

The Central Pacific of California was building the western end of the line during the same years. A young railroad engineer from the East, Theodore Dehone Judah (1826–63), was the first Californian seriously to consider the problem of building a railroad over and through the Sierras. He failed in his efforts to get a Pacific railroad act passed by Congress in 1859, but he did manage to enlist the support and money of four hard-headed merchants from Sacramento. Leland Stanford (1824–93), grocer and soon to be the first Republican governor of the state, Collis P. Huntington (1821–1900), and Mark Hopkins (1813–78), partners in a Sacramento hardware store, and Charles Crocker (1822–88), gold-miner turned dry-goods merchant, together incorporated the Central Pacific on June 28, 1861. Judah, chief engineer for the new company, after pushing his survey up the western slopes of the Sierras, returned to Washington to resume his lobbying for the Pacific railroad. Naturally, the congressional aid bill of 1862 named the Central Pacific as the builder of the western portion of the line. When Lincoln signed the measure, Judah sent a message to his western partners over the recently completed Pacific telegraph: "We have drawn the elephant. Now let us see if we can harness him up."

Harnessing the elephant was not easy. On January 8, 1863, the four merchants and their friends made speeches as they turned the first shovelful of earth on the muddy levee at Sacramento, but as construction got under way, Judah had the same trouble with his partners that Peter Dey was to have with the men of the Union

Pacific. The "Big Four" were happy to buy out Judah for $100,000, although they did give him the option of buying up each of their interests for the same amount. Before he had a chance to talk to the Vanderbilt group concerning these options, Judah was dead of yellow fever contracted on his trip east. The Big Four proceeded to build the road their way. Governor Stanford became president of the line, Huntington soon became the eastern agent for financial and political matters, Hopkins was made treasurer, and Crocker was put in charge of construction.

Labor was scarce in the West because most men were seeking their own fortune rather than a mere job. Crocker was having a hard time filling up his construction gangs until he thought of the Chinese. Few believed that the Chinese, who weighed, on the average, no more than 110 pounds, could stand up under the tough railroad work in the high Sierras. Crocker tried the first fifty as an experiment, was pleasantly surprised, and soon was hiring them by the hundred, importing many directly from Canton. Before 1867, six thousand Chinese laborers, queues covered by basket-hats, were making fills with hand barrows, sweating as they drilled away at Summit Tunnel, or were busy building snowsheds for protection against snow and avalanches. Eventually, the hard-working crews pushed the rails over and through the mountains. Earlier, in 1866, Congress had authorized the Central Pacific to build on through Nevada. In 1868, Crocker's construction gangs were hustling as fast across the flats of Nevada as the Union Pacific had moved over the plains of Nebraska. The Central Pacific had far less Indian trouble than the eastern line, partly because the tribes it encountered were less war-like than the Sioux, but also because Huntington was willing to give them a ride. "We gave the old chiefs a pass each, good on the passenger cars," he recalled, "and we told our men to let the Indians ride on the freight cars whenever they saw fit."

The rivalry of the two roads for extra mileage and extra government subsidy increased as Huntington worked on President Andrew Johnson (1808–75), while General Dodge, both as a congressman from Iowa and as an old army friend, sought the support of President-elect Ulysses S. Grant (1822–85). A compromise was reached on April 10, 1869, when Congress named Promontory Point as the meeting place. As the rails neared the Great Salt Lake (to the displeasure of Brigham Young [1801–77], not Salt Lake City) the rival companies laid track with new fervor. Crocker added some stalwart Irish to his docile but effective Chinese crews and proceeded to win a $10,000 wager from Union Pacific officials by laying ten miles of track in a single day.

On Monday morning, May 10, 1869, two special trains from California arrived at Promontory Point, bringing to the little tent construction town Leland Stanford, company officials, and their guests. The Union Pacific train was delayed until after noon because of floods and washouts on the line east of town. Finally arriving, Durant's shining Pullman brought not only the company's vice-president but also General Dodge, Sidney Dillon (1812–92), a director in both the Union Pacific and the Credit Mobilier, and the Casement brothers. Several companies of the 21st Infantry kept back the assorted crowd of perhaps five hundred Irish and Chinese workers, train crewmen, lesser officials, excursionists, and townspeople as the belated rail wedding finally got under way.

Several special spikes were presented, along with speeches of indorsement, before Stanford and Durant got down to the serious business of using a silver sledge to drive home the last spike into a polished tie of California laurel. Much to the delight of the professionals who had driven spikes all the way from Omaha to the Pacific, both men missed the golden spike the first time. As the spike was finally tapped into place telegraph operators reported the news to a waiting nation, the magnetic ball dropped from its pole on top of the Capitol dome in the nation's capital, a parade four miles long began to move in Chicago, and dozens of firebells rang in

San Francisco. The Central Pacific's share of the food and drink consumed in the festivities at Promontory Point came to $2,200. Bottles of champagne were quite prominent in the official picture of the two locomotives touching their cowcatchers over the golden spike. Perhaps because the picture seemed uncouth, but more likely because of the absence of a number of prominent figures, Leland Stanford commissioned an artist to recreate the scene in oil. The artist, although he crowded seventy important portraits into the massive panorama, was never able to satisfy the president of the Central Pacific. The American public was well satisfied with the actual event and took deep pride in the completion of a rail connection between the two oceans. The days of the Overland Mail stages and the Pony Express were clearly past.

The meeting of the Union Pacific (1,038 miles) and the Central Pacific (742 miles) in the spring of 1869 was followed by months and years of relocating, regrading, and reconstructing much of the original jerry-built line. It was also followed by revelations and scandals concerning the construction companies which had built the two lines. In the lush postwar period after 1865 Americans probably could not be expected to build 1,800 miles of railroad from Omaha to the Pacific without at least a trace of corruption. There was more than a trace. As the Ames brothers, Oakes (1804–73) and Oliver (1807–77), shovel-makers from Massachusetts, joined Sidney Dillon and Dr. Durant in thimblerigging the Credit Mobilier, none of the participants wished to be satisfied with a modest profit.

The Ames brothers were clearly not content to settle for the mere profit made from selling thousands of shovels to the Union Pacific. Oliver Ames succeeded General Dix as president of the company in 1868, and from his vantage point in Congress, Oakes Ames distributed shares of Credit Mobilier stock among those influential members of Congress where, as he put it, "they will do most good to us." The subsequent congressional investigation resulted in testimony which was to affect the careers of such men as Vice-President Schuyler Colfax (1823–85), Speaker James G. Blaine (1830–93), and James A. Garfield (1831–81). From the safety of their company offices the managers of the Union Pacific made millions through the operations of their construction company. In all they pocketed perhaps $23,000,000 in bonds, stocks, and cash and set up for their company a capitalization of $110,000,000, nearly half of which was pure water.

In the construction of the Central Pacific the Big Four were equally successful in making an extra, illicit profit for themselves. At first the road was built for the partners by a construction concern called Crocker and Company. The transparent connections between the Central Pacific and Crocker's false-front corporation were so obvious that in 1867 the partners shifted to dealing with a newly created Contract and Finance Company. Stanford, Hopkins, Crocker, and Huntington each subscribed equally to the new concern, and the four managed to get their accounts into such a shape that no outsider had a chance of understanding them. Any successful investigation was rendered quite improbable by the "accidental" destruction of the company's books, perhaps by fire, in the early seventies.

On the same day, July 2, 1864, that the Union Pacific had its federal land grant increased, a second transcontinental road was given a congressional charter and a land grant. The Northern Pacific Railroad Company was to build a line from a point on Lake Superior to one of the ports in the Pacific Northwest and to receive forty alternate sections of land for each mile completed within the territories and twenty sections for each mile built through organized states. Perhaps the project deserved the extraordinary promised land grant of forty million acres, for the terrain through which the road was to pass was so bleak that General Sherman described it "as bad as God ever made, or any one could scare up this side of Africa."

The company's first president, Josiah Perham (1803–68), Yankee showman and originator of the railroad-excursion idea, believed that the common people of the nation would flock to invest their savings in the new project. Perham failed in his dream of a people's railroad, and the road languished until Jay Cooke (1821–1905), Philadelphia banker, took over the financial management in 1869. The success of the project seemed assured as the top figure of Civil War finance took charge. Cooke sought $100,000,000 to finance the road's construction. The first rails were laid in Minnesota in February, 1870, and by the end of 1871 the line had reached the east bank of the Red River and the Dakota Territory. Nearly five hundred miles of line were completed by 1873, and the road was approaching a new station, Bismarck, North Dakota, named after the German chancellor, whose people were just starting to settle the prairie land. But the rapid construction had strained Cooke's resources, and the current Credit Mobilier scandal made Congress shy away from any program of additional aid. In September 1873, Jay Cooke and Company closed its doors, financial panic swept the country, and construction on the Northern Pacific came to a halt. Within a few months the line was in financial trouble, and George W. Cass (1810–88) was made receiver for the road.

Bismarck, the new town on the upper Missouri, remained the western terminus for the Northern Pacific until 1878. Meanwhile, Henry Villard (1835–1900), an immigrant from Bavaria and a well-known newspaper correspondent during the Civil War, had become interested in western railroads. By the late seventies Villard was in financial control of the Oregon Steamship Company and the Oregon Central Railroad, which together fairly well monopolized transportation in the Far Northwest. Villard soon was looking eastward toward the Northern Pacific, which was beginning to show signs of life. He organized the "Blind Pool" among his friends and New York investors, who on faith subscribed

$8,000,000 to carry out an unannounced project. Villard used the funds to buy control of the Northern Pacific, with himself as president. As Villard pushed construction vigorously on both main and branch lines, he used as many as 25,000 men, more than half of whom were Chinese. On September 8, 1883, west of Helena, the last rails were laid in the presence of such distinguished guests as President Chester A. Arthur (1830–86), General Grant, and James Bryce (1838–1922) of England. Through service to the Pacific was now possible because of a western connection with the Oregon Railway. The Northern Pacific completed its own line over and through the Cascade Range to Seattle at a time when there were still no states between Minnesota and Oregon. Also in the middle eighties, the Union Pacific, not to lose out in the trade of the Pacific Northwest, projected and built the Oregon Short Line from southwestern Wyoming into western Oregon.

While the Northern Pacific was experiencing its first siege of bankruptcy and receivership and before the memory of the Credit Mobilier became blurred, another northern transcontinental line was beginning to take shape. In 1878, James Jerome Hill (1838–1916) and his Scotch Canadian associates, Norman Kittson, Donald Smith, and George Stephens, secured control of the bankrupt and rusting St. Paul and Pacific Railroad. One-eyed Jim Hill had come to St. Paul from his native Ontario in 1856, intent upon traveling west with some band of Pacific-bound fur trappers. Missing the last brigade of trappers by a few days, the energetic and observant youth stayed on in St. Paul, soon to become completely involved in the business of the city that was to be his base of operations for sixty years. Hill had twenty years of frontier freighting, merchandising, and transportation experience behind him when he and his banking and Hudson's Bay Company friends from Canada took over the ailing St. Paul and Pacific.

The first move of the new management was to extend the line northward toward Winnipeg and a connection with a branch of the Canadian Pacific. Hill next pushed west. Reorganized as the St. Paul, Minneapolis, and Manitoba, or "Hill's Folly" to his jeering critics, the line reached Great Falls, Montana Territory, by 1887. Hill built his line well, added branches where he saw profitable traffic, and was willing to sell a land-seeker or a healthy immigrant farmer cheap passage in a second-class car if the farmer would agree to homestead near the railroad. The completed road, since 1889 called the Great Northern Railway, reached Seattle in July 1893. The Panic of 1893 stopped the giant celebration that had been planned, but "Empire Builder" Jim Hill had the satisfaction of seeing all the other transcontinental lines face receivership while his own road continued regular dividends throughout the depression nineties. The dividends were clearly the result of Hill's insistence upon careful original construction (meaning easier curves and lower grades), a conservative financial structure, and diligence in management. The operating ratio of the Great Northern normally was six to eight percentage points below that of the other Pacific roads. The financial difficulties of the nearest competitor, the Northern Pacific, were such that the Hill interests were in effective control of this line by the turn of the century. Nor did the depression nineties slow construction on the parent line. Between 1891 and 1907, the year Hill retired from its presidency, the Great Northern built an average of one new mile of road for every working day of the year. Even as he neared retirement Hill continued his varied interests in attracting immigrants to the Northwest, improving agricultural production in the area, and capturing a greater share in the Oriental trade.

In the same years that Cooke, Villard, and Hill were laying track across the northern plains, other promoters and builders were busy in the Southwest. Cyrus K. Holliday (1826–1900), "Free Soiler" and founder of Topeka, Kansas, managed to get a state charter for his Atchison, Topeka, and Santa Fe in 1859 and a modest three-million-acre land grant from Congress in 1863. Twenty-eight miles of track were laid in 1869, and Dodge City (soon to be the most famous cowtown of the West) was reached two years later. In 1872 the line had been built to the Colorado border, assuring the government land grant. Since the Santa Fe crossed the major cattle trails farther south than did rival roads, much of the early prosperity of Holliday's line was clearly caused by the expanding cattle business. As the Santa Fe neared Colorado, business interests in Denver, still disgruntled at being left off the main line of the Union Pacific and only partially pacified with the recently completed Kansas Pacific (finished to Denver in 1870), were disturbed at the prospect of losing southern Colorado business to the new rival. In 1870 the Denver people chartered the Denver and Rio Grande to challenge the Santa Fe. General William Jackson Palmer (1836–1909), who had served a railroad apprenticeship as private secretary to J. Edgar Thomson before the Civil War made him forget his Quaker training, was the first president of the new Colorado line.

The Panic of 1873 slowed the competition between the two roads, but later in the decade open warfare broke out as both lines sought to possess strategic mountain passes. William B. Strong (1837–1914), now general manager of the Santa Fe, beat General Palmer's construction gangs to Raton Pass, the best and easiest gateway into New Mexico, but the Denver and Rio Grande ultimately won out in the scramble for narrow Royal Gorge west of Pueblo. Palmer finally pushed the mountain line through to Ogden by 1882. William Strong, president of the Santa Fe since 1881, rapidly expanded his own system through a combination of leases, purchases, and construction. By the time he retired in 1889 he was operating a seven-thousand-mile system stretching from Chicago to the Gulf and the Pacific. A distinctive feature of the growing Santa Fe was the restaurant service provided by Frederick H. Harvey (1835–1901) after 1876. Before the turn of the century hundreds of Harvey Girls ("Young women of good character, attractive and intelligent, 18 to 30") in dozens of Harvey Houses were serving some of the best meals in the Southwest. Entirely apocryphal is the story that Fred Harvey's dying words were: "Slice the ham thinner."

The remaining major nineteenth-century Pacific railroad followed a route, important in the earliest plans to reach the West by rail, running through the entire length of the Gadsden Purchase. After completing their main line to Promontory Point, the Big Four of the Central Pacific soon had ambitions to build a major branch into southern California. Incorporated as the Southern Pacific, for the sake of a more favorable public opinion, and using the Western Development Company as the device to award themselves the construction contracts, Stanford, Huntington, Hopkins, and Crocker pushed the new line down to Yuma, Arizona, by 1877. East of the Colorado River construction was held up by the Yuma Indian Reservation and considerable vacillation by the Secretary of War and army officers resident on the reservation. Tactful interviews by Huntington with President Rutherford B. Hayes (1822–93) and members of his cabinet finally resulted, in early October, 1877, in an executive order which legalized construction into Arizona.

In the same years, Thomas A. Scott (1823–81) was building the Texas and Pacific Railroad from Shreveport, Louisiana, across the central plains of Texas in an effort to reach the congressional land grant available to his road in New Mexico and Arizona. In the early seventies Scott was a giant in railroad management. In addition to being president of the Texas and Pacific, he was vice-president of the Pennsylvania, was the prime mover in the latter road's efforts to build a rail empire in the South, and from 1871 to 1872 was president of the Union Pacific. The success of the Southern Pacific in crossing Arizona convinced Scott that he could never successfully claim the land grant west of El Paso. In 1880 he was glad to sell his holdings in the Texas and Pacific to Jay Gould (1836–92), who had earlier purchased his Union Pacific securities. When Gould's Texas and Pacific made a junction with the Southern Pacific in 1882 a short distance east of El Paso, another transcontinental rail route was complete. Not content with this joint line, the Big Four soon completed arrangements to buy and build a route of their own east of El Paso via San Antonio and Houston. This consolidation and construction gave them through service to New Orleans by 1883.

Within a generation after the Civil War five transcontinental roads had built through to the Pacific Coast. Their construction involved many unprecedented and complex problems of engineering and finance. The promoters, politicians, financiers, and contractors who planned, financed, and built the lines lived in a lusty expansive America whose code of business ethics was not high. In the years after Appomattox most Americans with money regarded the Pacific-railroad proposals more as wildcat schemes than sure-fire bonanzas. Even the heavy government subsidies frequently were not sufficient to tempt the investor. Many bankers saw better places for their investment funds back East, where there were some people. The false-front construction company opened up the possibility of an extraordinary profit, which the investors felt the risky railroad ventures deserved. Even though their business ethics were low, the completed Pacific railroads were genuine accomplishments which hastened the economic expansion of the entire nation.

The main stems of the five Pacific railroads were obviously only a small portion of the total western rail construction in the generation after the Civil War. If one arbitrarily defines the West as the twenty-two states west of the Mississippi River (including Minnesota and Louisiana), the total rail mileage by decades is as follows: 1865, 3,000 miles; 1870, 12,000 miles; 1880, 32,000 miles; 1890, 72,000 miles; 1900, 87,000 miles; 1910, 118,000 miles; and 1920, 127,000 miles. Western rail mileage more than doubled in each decade up to 1890, expanded more rapidly than eastern mileage throughout the period, and by 1920 (when the national total [253,000 miles] had just passed its all-time peak) could claim slightly more than half the national total.

Long before the turn of the century it became obvious that the railroad map (and service) of the prairie states was quite different from that of the mountain states farther west. Between the Mississippi River and the ninety-eighth or one hundredth meridian the railroads of the area formed a network criss-crossing in every direction. To the west, in the high plains and mountain area, the lines were fewer in number, and most of them were east-west roads, with very few running north and south. The eleven prairie states west of the Mississippi and mainly east of the hundredth meridian (in two tiers, Minnesota to Louisiana and North Dakota to Texas) could always claim from two to three times the rail mileage found in the eleven mountain and coastal states to the west. By 1920 only one of the prairie states (South Dakota) had less than five thousand miles of railroad, and four of the eleven (Texas, Iowa, Kansas, and Minnesota) could claim over nine thousand miles of line. These last four states ranked first, fourth, fifth, and sixth in mileage among the forty-eight states in 1920. In contrast, among the mountain and coastal states only four (Montana, Colorado, Washington, and California) had as much as five thousand miles of road, and none of the eleven had nine thousand miles of trackage. By the end of World War I the prairie region, with a small third of the national area, possessed a large third of the nation's rail mileage, while the eleven mountain and coastal states, with more than 40 per cent of the national area, had but a sixth of the nation's total rail mileage.

In addition to the Pacific railroads, which served several of the prairie states, four major "Granger" lines—the Chicago, Burlington & Quincy; the Chicago, Milwaukee & St. Paul; the Chicago & North Western; and the Chicago, Rock Island & Pacific—

MAJOR
TRANS-MISSISSIPPI
and PACIFIC RAILROADS

Line	Completed by
1 Union Pacific	1869
2 Central Pacific	1869
3 Kansas Pacific	1870
4 Denver Pacific	1871
5 Oregon Short Line	1882
6 Oregon Ry. & Nav.	1883-84
7 Northern Pacific	1883
8 Denver & Rio Grande	1882
9 Texas & Pacific	1883
10 Southern Pacific	1883
11 Santa Fe	
11a Atlantic & Pacific	
11b Extension to Chicago	1888
12 Great Northern	1893

were important in the northern central plains region. The Granger area in the decades after the Civil War might be described as the nine states from Illinois, Missouri, and Kansas north to Canada. Seven of these grain-growing states were west of the Mississippi, while two, Illinois and Wisconsin, were to the east. All were economically subservient to Chicago and, to a lesser degree, St. Louis and the Twin Cities. Each of the four Granger railroads had Chicago as its main eastern terminal, and each of the four served at least seven of the nine states. Their total mileage in 1890, some 18,000 miles, was over a third of the total for the Granger territory. With the exception of the shorter Rock Island line, each of the roads in the late eighties was grossing annual revenues of from $23,000,000 to $27,000,000 for carrying the pork, beef, and grain of the region to the stockyards and elevators of Chicago. All of the lines paid regular dividends.

Each of the lines, or predecessor companies, was organized at mid-century, and three of the four were built to serve Chicago. William Ogden had started the Galena and Chicago Union in 1848 and was also the first president of the Chicago & North Western when it took over in 1859. While English capital was important in the financial management of the North Western in the nineteenth century, it was not this fact but, rather, the original location of the Galena and Chicago Union depots which dictated that the road's double-track line should be operated left-handed. For seventy-five years the North Western's main line from Chicago to Omaha provided through passenger service to the coast in connection with the Union Pacific. The Milwaukee took over this service in 1955. The Chicago, Burlington & Quincy was the second of the Chicago roads, the Aurora Branch Railroad receiving its original charter on February 12, 1849. John Murray Forbes, already deeply interested in Chicago because of his Michigan Central, wished to see the Burlington push on into Iowa, Missouri, and Nebraska. His dream was a reality by the time he became president of the road in the late seventies. Under his successor, Charles El-liott Perkins (1840–1907), who was president of the road from 1881 to 1901, the Burlington reached Denver in 1882, St. Paul in 1886, and Billings, Montana, in 1894. The road was noted for its careful financial management under both Forbes and Perkins.

Low fixed charges and prudent management were also typical of the smaller Chicago, Rock Island & Pacific after the Civil War. Dividends of 7 or 8 per cent were normal until the late eighties, when extensions into Kansas and Colorado reduced the average earnings of the road somewhat. The Rock Island was one of the first western lines to experiment with the new steel rails after the Civil War. By 1880 it could boast of being the only road to have an all-steel track from Chicago to the Missouri River. The last of the four major Granger roads, the Chicago, Milwaukee & St. Paul, started out in 1850 as a short line west from Milwaukee. The pro-moters of the original line, the Milwaukee and Mississippi, hoped that their project would permit Milwaukee to rival Chicago as a lake port and railroad center. Under the presidency of Alexander Mitchell (1817–87), Milwaukee banker and insurance man, the road was extended to St. Paul shortly after the Civil War and had its own line into Chicago early in 1873. By 1890 the Chicago, Mil-waukee & St. Paul was the largest of the Granger roads, with branches extending to Omaha and into North and South Dakota. Still ambitious, the road added "Pacific" to its name, built to the coast between 1906 and 1909, and electrified more than four hun-dred miles of its mountain line. The last of the transcontinentals was never prosperous. Stiff competition from the Hill roads, a top-heavy financial structure, and bad times in the Northwest forced the line into bankruptcy in 1925.

Four of the first five Pacific Railroads and all of the Granger lines just mentioned received substantial land grants from the United States government. Between 1850 and 1871 most western railroads sought, and generally received, land grants from the federal government. When Senators Douglas and King of Illinois and Alabama urged the Illinois Central–Mobile & Ohio land grant in 1850, eastern as well as western senators favored the idea. William H. Seward, in Senate debate on this first land-grant measure, contended that the best interests of Americans would be served if their western lands were brought "into cultivation and settlement in the shortest space of time and under the most favorable auspices." Land grants made in the fifties were modest in amount, normally being six sections per mile of railroad, with the grants being given to the states. After the first grant in 1850 and before the Civil War, nearly twenty million acres were granted to the first tier of trans-Mississippi states plus Wisconsin, Michigan, and Florida. One of the most extensive grants went to Iowa in 1856 for the building of four east-west lines across the state.

The Civil War and postwar grants were even more generous. Most of the Pacific railroads received twenty sections of land per mile of road, and the Northern Pacific was offered forty sections for each mile built in the territories. The transcontinental grants were frequently given directly to the railroad rather than to the states. By the time the last important land grant was given to the Texas and Pacific in an act of March 3, 1871, grants of land totaling more than 170,000,000 acres had been made available to eighty-odd railroads. Since nearly half the projected railroads were never built or completed in time to earn the grants, some 35,000,000 acres were later forfeited and returned to the federal government. Some additional land claims were released by several railroads under the Transportation Act of 1940.

By 1943 the railroads of the nation had received full and final title to 131,350,534 acres. Almost 90 per cent of this acreage was located in twenty states west of the Mississippi (Texas, having retained control of her land, received no federal land grants, nor were there any in the Indian Territory), but 15,436,000 acres were in the six more easterly states of Alabama, Florida, Illinois, Michigan, Mississippi, and Wisconsin. The grants in the several states varied greatly in size, ranging from an insignificant amount in South Dakota to 10,697,490 acres in North Dakota (23 per cent of the state's area) and 14,736,919 acres in Montana (15 per cent) The total grant amounted to 9.5 percent of the area of the twenty-six states and to 6.8 per cent of the total land area of the nation. It was an area greater in acreage than the total of the perfected homestead entries in the first half-century of the Homestead Act (1862–1912) and an empire four-fifths as large as the five states in the Old Northwest. In addition to federal assistance, the railroads also received a number of land grants from nine states amounting to another 48,883,372 acres. The federal land grants aided in the construction of 18,738 miles of line, a figure that is less than 8 per cent of the current total mileage in the nation but one that amounts to 20 per cent of the national total of 93,000 miles in 1880 (by which time most of the land grants had been claimed) and to 40 per cent of the 1880 mileage of the twenty-six states in which the land grants were located.

There has been considerable argument about the value of the land grants received by the railroads. If the cost to the federal government, 23.3 cents an acre (purchase price, payment to Indian tribes, and surveying costs), is used as a basis, the 131,000,000 acres would be valued at only $31,000,000, a figure that is ridiculously low. The standard price asked by the government for its land during the land-grant period was $1.25 an acre, but actual sales in the period 1850–71 show an average price of only 97 cents an acre. These figures also give us a total value that is too low. Perhaps it would be more reasonable to ask how much the railroads gained from the sale of their land grants. A government study made in 1941 reported that the railroads received on an average $3.38 an acre for their land. Much of the land in the desert West was certainly worth very little, but in the seventies the Union Pacific was selling its Nebraska grants at $3.00 to $5.00 an acre, and the Rock

Island was selling its Iowa land grants at an average of $7.00 to $8.00 an acre. In the same years the Burlington was getting from $11.00 to $14.00 for its Iowa acres and from $4.00 to $8.00 for land farther west in Nebraska. Of course from these figures should be subtracted the extensive promotional and advertising costs incurred by the railroads. Thus a fair estimate of the value of the land grants to the railroads would be about $500,000,000.

Most of the congressional land-grant acts followed the wording of earlier wagon and canal acts in requiring that the land-grant railroad should "be and remain a public highway for the use of the government of the United States, free from toll or other charge upon the transportation of any property or troops of the United States." As subsequently determined by the courts, this amounted to a reduction of 50 percent from the ordinary rates for government shipments. Noting the continued movement of troops and government freight across Iowa in 1859, Charles Russell Lowell (1835–64), young nephew of the poet and in charge of the Burlington land office in Iowa, wrote: "It may be found that even with the most liberal construction of the grant, the Government has not been so 'munificent' as sharp." Sharp or not, a congressional report in 1945 estimated that in the years since 1850 the federal government had saved $900,000,000 because of the reduced rates. True, much of the $900,000,000 savings was made at the expense of non-land-grant roads, since many of these lines had been willing, through "equalization agreements," to take the government traffic at the same low rates available on the land-grant lines. It would seem reasonable, however, that perhaps half the government savings were made on land-grant line traffic, giving a figure roughly equal to the amount which the railroads received for their land. The rate reduction on land-grant roads was finally terminated with legislation which became effective on October 1, 1946.

The major contribution of the land grants to western railroads was their furnishing a basis of credit so that building could be started. Most railroads obtaining grants mortgaged their land long before they completed final certification with the government, obtained the patents to the land, or sold it on long-term credit to settlers. The typical railroad was still selling its land years after the line itself had been completed.

Important in any consideration of the railroad land grant is the general public's impression of the extent of such grants. This impression has been much exaggerated from the eighties until recent years because of some faulty mapwork. In the 1884 election the Democratic party, thinking to embarrass the Republicans, who had been quite generous with land grants, issued a campaign poster consisting of a map with this caption: "How The Public Domain Has Been Squandered—Map showing the 139,403,026 acres of the people's land ... worth at $2 an acre $278,806,052 given by Republican Congresses to Railroad Corporations." The map was correct only if one remembered two important facts: (1) only alternate sections of land had been granted—creating a "checkerboard" pattern which the map could not show; and (2) the shaded outer limits of the grants included indemnity lands (where the railroad could have substitute land only if their sections in the primary strip had been previously taken).

These important qualifications were not mentioned on the map, and as a result it exaggerated the extent of the grants roughly fourfold. Iowa was shown 90 per cent black when in reality the total grants amounted only to 13 per cent of the area of the state; Kansas was shown as 62 per cent instead of a correct 16 per cent; California was 40 per cent instead of 12 per cent; Michigan 75 per cent instead of 9 per cent; and Minnesota 70 per cent instead of 19 per cent. This original map was widely copied and recopied by dozens of high school and college history textbooks for nearly sixty years. Only since the middle 1940s has a basic correction been made. For two full generations Americans grew up thinking that nearly all of Iowa had been given to the railroads. Nor was the average student informed that in the final settlement of 1898–99 the government received $63,023,512 in principal plus $104,722,978 in interest in repayment for the original bond aid of $64,623,512 extended to the Union Pacific–Central Pacific and four branch-line companies just after the Civil War.

The railroads were far from faultless. In Nebraska the Union Pacific avoided local taxation on its millions of acres for years, by being intentionally slow and tardy in applying for patents to its grants. Many other roads followed the same practice. Also, many railroad companies and favored company personnel made substantial corporate or personal profit through the sale of land-grant acreage by subsidiaries or affiliates. These profits are not included in the government estimates of the value of land sales made by the several railroads. One final evil was the injustice to would-be settlers brought about by the government practice of prohibiting all settlement in a region until the railroad had received its full land grant. These prohibited areas in some states did approximate the shaded areas shown on the original map of 1884. When the homesteaders violated the rule and settled on land claimed by the railroad, the result was often endless litigation and sometimes violence. The problem was partially resolved in 1887 when President Grover Cleveland (1837–1908) opened to the ordinary settler much of the land within the railroad indemnity limits.

In the nineteenth century Americans experienced a succession of frontiers as the explorer, trapper, cowboy, miner, and finally the farmer conquered the West. Between the Civil War and World War I western railroads participated fully in the last three of these western developments. Perhaps the first was the role played by the expanding rail lines in the frontier cattle industry between 1867 and the eighties. The Kansas Pacific was building west across Kansas in 1866, the same year that Texas cattle drivers met the triple frustrations of hostile farmers, timbered terrain, and bandit rustlers in their drive toward Sedalia, Missouri.

Joseph G. McCoy (1837–1915), a livestock shipper from Illinois, believed that there should be some intersecting point of cattle trail and railroad in Kansas where northern and eastern buyers could meet Texas cattlemen with mutual profit. Shortly after the president of the Missouri Pacific had rudely ordered him out of his office, McCoy signed a favorable contract with the Hannibal and St. Joe Railroad to ship cattle from Missouri to Chicago. In searching for a trail head in central Kansas, McCoy found that neither Salina nor Solomon City would tolerate the idea of stockyards. He turned to Abilene on the Kansas Pacific, a town so small that only one of the dozen log huts could claim a shingle roof and a place so poor that the saloon keeper supplemented his income by raising prairie dogs. McCoy built his stockyards and shipped out the first train of twenty cars of longhorn cattle on September 5, 1867. By the end of the year a thousand carloads had been sent to market in Chicago. In 1868 some 75,000 Texas cattle were sold in Abilene, 300,000 head came north in 1870, and the peak was reached with 700,000 in 1871. None of the 75 bartenders (three eight-hour shifts of 25 men each working seven days a week) at the Alamo Saloon in Abilene now had any time for prairie dogs.

As the railroads pushed farther south and west and as the hordes of settlers and farmers followed them, the cattle trails and the cowtowns were pushed ever westward. Newton and Dodge City on the Santa Fe, Ellis farther west on the Kansas Pacific, and Ogallala on the Union Pacific in Nebraska all increasingly shared in the cattle business. Ellis and Dodge City together shipped out more than a million cattle between 1876 and 1879, with the latter place, being farther south, getting the bulk of the traffic. Between four and five million cattle had been driven to northern markets by 1880, but the decline in the traffic was well established by the latter year. Farmers were planting wheat on the old buffalo ranges, barbed wire was creeping ever farther to the west, and both Kansas and Colorado were soon to have effective quarantine laws against Texas cattle.

New railroad construction from the North into Texas was also making the long drive unnecessary. The Missouri-Kansas-Texas Railroad reached the Red River and northern Texas in 1873, and the St. Louis, Iron Mountain, and Southern Railroad diagonally crossed Arkansas to Texarkana shortly thereafter. Texas had more than quadrupled her own rail mileage in the seventies, increasing it from 711 miles in 1870 to 3,244 miles in 1880. Clearly, the days of the open range and the long trail drive were about over.

Also during the decade of the seventies a second frontier, that of the Rocky Mountain miner and prospector, was helped by the western railroad. A mania for narrow-gauge railroads appeared in the United States, as in England, shortly after a paper, "The Gauge for the 'Railways of the Future,'" was read by Robert F. Fairlie (1831–85) in 1870 before the annual meeting of the British Railway Association. Fairlie and his adherents argued very plausibly that a road of 3-foot 6-inch gauge would be much cheaper to build, equip, and maintain than standard-gauge line. It was also pointed out that a narrow-gauge line, with sharper curves and lighter equipment, was well adapted to mountainous regions or could be used in areas where the expected business was light.

The advocates of the lighter narrow-gauge roads soon started to build them in the United States. Many companies followed the lead of General Palmer's three-foot-gauge Denver and Rio Grande, and by 1874 nearly 1,700 miles of the new type of line had been built. By 1878 the total was 2,862 miles, and the 1880 trackage figures showed that 5,200 miles (or more than 5 per cent of the national total) were of the new gauge. While 2,000 miles of narrow-gauge were built in 1882, the popularity soon started to decline as many roads subsequently shifted to standard gauge or at least added a third outside rail (becoming double gauge) to accommodate standard-gauge equipment.

While much of the narrow-gauge trackage in 1880 was in eastern states, with Ohio, Pennsylvania, and Illinois each having over 200 miles, the new type of line never seriously challenged standard gauge in the eastern part of the country. But in the eleven western mountain and coastal states the narrow trackage of 1,200 miles was nearly a sixth of the total western mileage. California had nine very short narrow-gauge roads, while in Colorado the new type of line amounted to almost a third of the total. Many of the new lines served mountain mining regions, especially those in California, Nevada, and Colorado. When General Palmer found his ambitions for a southern line into New Mexico and Mexico thwarted by the Santa Fe possession of Raton Pass, he soon turned his three-foot line into the mountains west of Pueblo.

After silver was discovered in vast amounts at Leadville in the late seventies, Palmer turned his line in that direction. John Evans (1814–96), early Colorado territorial governor, co-founder of two Methodist colleges, and former president of the Denver Pacific (Denver to Cheyenne), also wanted his narrow-gauge Denver, South Park, and Pacific to serve Leadville. Palmer's Denver and Rio Grande reached the booming silver town first. Former President Grant, just returning from his world tour, was aboard Palmer's private car "Nomad" as the first train entered the city to be welcomed by twelve brass bands and Horace A. W. ("Haw") Tabor (1830–99), bonanza king and lieutenant governor of the state. The rails of the Denver, South Park, and Pacific later reached the city, but with a grade so steep that on one occasion when a circus train was headed for Leadville, the elephants were unloaded to help push the train into the station. Soon, branches of the Denver and Rio Grande were snaking their way into Aspen, Durango, Silverton, and Farmington across the New Mexican border. There was economy in building the narrow-gauge lines, but the prosperous patrons of the little roads demanded, and received, service of the finest sort in the small-scale mahogany and red plush parlor, dining, and Pullman cars.

A comparable story of gold and silver strikes, of boom towns and the narrow-gauge lines that served them, could be found in both California and Nevada. In central Nevada in the middle seventies the Eureka and Palisade Railroad pushed its three-foot track 90 miles south of the main line of the Central Pacific to the bonanza claims of both silver and lead waiting in Eureka. A few miles west and three years later, in 1879, a parallel line, the Nevada Central Railway, was headed for the silver mines of Austin. General James H. Ledlie (1832–82), who had been the contractor for all the bridges, trestles, and snowsheds on the Union Pacific, completed the 93-mile Nevada Central just before midnight on February 9, 1880, the last day that the road was eligible for a Lander County subsidy of $200,000. Even then the deadline had been met only because the Austin Common Council had extended the limits of their town half a mile toward the narrow-gauge line.

In the years just after the Civil War western railroads played a major role in the frontier of the cowboy and the cattle industry of the Great Plains. In the same period an equally important contribution had been made by the narrow-gauge line in the expansion of the mountain mining frontier. The first full post-war generation (1865–1900) saw a vastly more important contribution to prairie agriculture by the railroad.

As prairie farming expanded in the generation after the war, American agriculture experienced a revolution composed of three factors: (1) an expansion and westward shift of the farming domain; (2) the use of new farming techniques (dry farming, irrigation, and increased mechanization) to meet the needs of prairie farming; and (3) the increased use of the railroad to distribute and market the western farm product. The expansion of the agricultural area can be seen in the nearly threefold increase (2,044,000 to 5,737,000) in the number of farms between 1860 and 1900 and in the fact that more new farm land was brought under cultivation in the generation after the Civil War than in the entire previous history of the nation.

In the same years the center of crop production shifted westward. Whereas Illinois and Indiana were first and second in wheat production in 1859, this honor had shifted to Minnesota and North Dakota by 1899. Illinois did retain its first-place rank in corn production in the same forty years, but second, third, and fourth places shifted, respectively, from Ohio, Missouri, and Indiana to Iowa, Kansas, and Nebraska. Mississippi had been first in the production of cotton in 1860, but by the turn of the century Texas had become the center of the Cotton Kingdom. The availability of transportation was an obvious necessity as this shift in the centers of agrarian production occurred. In the six leading prairie farm states mentioned above (Minnesota, North Dakota, Iowa, Kansas, Nebraska, and Texas) total railroad mileage increased from less than 1,000 miles in 1860 to over 42,000 miles in 1900. The availability of transportation in the western farming region helped create a shift from subsistence to commercial farming and made agriculture an intimate though subordinate factor in the post–Civil War industrial system.

Prairie farming was specifically tied into the industrial system in its nourishment of and dependence upon a whole series of cities on the eastern edges of the prairie. Chicago had its wheat pit, stockyards, and packing plants. Kansas City, Omaha, East St. Louis, and Milwaukee also had expanding facilities to serve the livestock farmers of the West. A trio of cities on the upper middle Mississippi, Rock Island, Moline, and Davenport, held a position in the production of farm implements surpassed only by the production of Chicago. The flour mills of Minneapolis served the western farming states and the nation. This entire agricultural-industrial complex was served by the growing rail network basically radiating out of Chicago through such rival but subordinate focal points as St. Louis, Kansas City, Omaha, and the Twin Cities. Admittedly, these prairie railroads were often guilty—as were most railroads in the late nineteenth century—of many abuses. Rate wars, pooling arrangements, rebating, discriminations of the long-

and-short-haul, railroad-owned elevators that downgraded the farmer's grain—all these things plagued and irritated the farmer of the prairie.

Western railroad rates in the years after the Civil War were not low. In the years between 1866 and 1870 the four major Granger lines (Burlington, North Western, Milwaukee, and Rock Island) plus the Illinois Central had average freight rates ranging from 2.2 cents to 2.5 cents per ton-mile for the region from Chicago to the Missouri River. East of Chicago in the same years, the Lake Shore & Michigan Southern, the Pennsylvania, and the Pittsburgh, Fort Wayne, and Chicago had average freight rates ranging from 1.25 cents to 1.6 cents per ton-mile. That this same rate advantage for movements east of Chicago continued until 1900 was explained in part by the heavy volume of freight traffic in the trunk-line region between Chicago and the Atlantic coast. West of the Missouri, railroad freight rates were still higher. East of the Missouri in 1880 the four Granger lines and the Illinois Central had average freight rates of 1.4 cents a ton-mile, while to the west the average for the Burlington, the Santa Fe, and the Union Pacific was 2.5 cents a ton-mile. In 1890 the traffic east of the Missouri had an advantage of 0.95 cents against 1.25 cents a ton-mile, and in 1900 the eastern area still retained this advantage by a margin of 0.83 cents to 1.03 cents a ton-mile.

Railroad freight rates seemed high to the western farmer in the generation after the Civil War, but they were, in fact, dropping more rapidly than the price of farm products, as can be seen in table 4.1. Farm prices dropped only 37 per cent from 1870 to 1900, but freight rates dropped nearly 70 per cent in the thirty-year period and 59 per cent west of the Missouri in the twenty years from 1880 to 1900.

Distance was as important a factor in the farmer's transportation problem as high freight rates. The West was a big place; its territories and states dwarfed most eastern states, and, naturally, it

TABLE 4.1

RAILROAD FREIGHT RATES FROM 1870–1900

Year	Price of Farm Products (1910–14 = 100)	Three Railroads East of Chicago (cents per ton-mile)	Five Railroads Chicago to Missouri River (cents per ton-mile)	Three Railroads West of Missouri (cents per ton-mile)
1870	112	1.26	2.26	...
1880	80	.81	1.40	2.50
1890	71	.66	.95	1.25
1900	71	.53	.83	1.03

was a long way to market. To get his produce to market the farmer in South Dakota or Kansas might have to send it several times the distance required by his competitor in Ohio or Indiana. The average length of all railroad freight movements in seven northwestern states (Iowa, Minnesota, Nebraska, North and South Dakota, Wyoming, and Montana) in 1890 was 201 miles, as contrasted to a national average of 113 miles and a 69-mile average haul in the New England states. High basic freight rates plus distance resulted in transportation costs to the prairie farmer that placed him at a significant economic disadvantage.

In spite of high freight rates, the long haul to market, and other transportation discriminations, farmers and settlers poured into the prairie states after the Civil War. Veterans headed for western homesteads via the new railroads of Iowa and Missouri. Second-class cars for immigrants and land-seekers were crowded in most of the prairie states in the seventies and eighties. The land agents of the Burlington, the Northern Pacific, and the Santa Fe all competed with each other and with the rival attractions of the steel mills and mines of Ohio and Pennsylvania as they met the immigrants and steerage passengers at eastern seaports. The extensive colonization literature of the Santa Fe paid off in 1874 when company agents convinced 1,900 Russian Mennonites, complete with

$2,000,000 in gold drafts, that their future lay in fertile plains of Kansas. The Northern Pacific gained as settlers tried to emulate the reported success of Oliver Dalrymple's bonanza wheat farm in the Red River Valley of the Dakota Territory. The Burlington promoted extensive programs in agricultural research and new dry-farming methods in western Nebraska and eastern Colorado, even though it had no land to sell in those areas.

Settlers came in increasing numbers. The annual average of perfected homestead entries rose to more than 2,500,000 acres in the eighties and to more than 3,000,000 acres in the next decade. The first tier of trans-Mississippi states (Minnesota to Louisiana) more than doubled in population, from 4,500,000 to nearly 10,000,000, between 1870 and 1900. In the second tier of states (North Dakota to Texas) the increase during the three decades was more than fivefold: from 1,300,000 to over 7,000,000. The population of Nebraska increased nearly nine times between 1870 and 1890, and that of the Dakotas shot up from 14,000 to more than 500,000 in the same years. As the frontier receded, the western railroad network grew more rapidly than the population. Between 1870 and 1900 the rail mileage of the eleven trans-Mississippi prairie states had increased nearly eightfold, while the population had only tripled. This trend toward a relatively greater rail service was a national one, since the miles of line per million inhabitants in the country had increased from 985 in 1860 to 1,858 miles in 1880 and to 2,544 miles in 1900.

The settling of America's last frontier, the trans-Mississippi and mountain West, and the rapid growth of the nation's railroads were in many ways simultaneous developments. As the early settlers pushed into the Great Plains just after the Civil War, they were immediately helped in their westward movement by a rail network that had already reached the fringes of the frontier. A long generation later, the frontier was fully closed as statehood came to the last of the forty-eight continental states (Oklahoma in 1907, Arizona and New Mexico in 1912). In the same years, the railroad was nearing the close of its golden age as new, competitive, and aggressive transportation facilities began to appear. But in the short half-century between Appomatox and World War I the railroads had helped the pioneer as he faced a prairie frontier that was unique in its lack of wood and water. They had built, with generous aid from the government, not one but many lines across the mountains to the Pacific. The railroad had speeded Texas-raised cattle from the Kansas cowtown to the Chicago stockyards, had made more accessible the riches of the silver and gold mines of Colorado and Nevada, and had taken to market the crops of the western settler, who was now a commercial farmer. True, the railroad corporations and the hard-headed, hard-hearted men who ran them were often guilty of charging ruinous freight rates and making other discriminations which seemed inexcusable to farmers beset by falling farm prices, high interest rates, and the necessity of competing in a world market. But it must be said that the western railroads had done much to hasten the settling and closing of America's last frontier.

5

Corruption, Discrimination, and Regulation

The United States experienced a record growth in both population and industrial development in the half-century after the Civil War. American railroads easily kept pace with this expansion. Not only the West but the entire country expanded in population and strength in the years between 1865 and the nation's entrance into World War I. As America shifted from an agrarian people to an industrial nation, the population nearly tripled, rising from 35,700,000 in 1865 to 103,400,000 in 1917. In the same half-century the rail network increased more than sevenfold, from 35,085 miles in 1865 to an all-time high of 254,037 in 1916. By the second decade of the twentieth century the only significant portions of the United States more than twenty-five miles from rail service were in the relatively unpopulated plateau and mountain areas of the West. Almost the entire nation was literally within the sound of a locomotive whistle. In the five decades between the wars the volume of the railroads' gross operating revenue increased at least twelvefold, from perhaps $300,000,000 in 1865 to over $4,000,000,000 in 1917. The rail lines in the middle teens still possessed a fairly complete monopoly of the nation's commercial freight and passenger movement. The new transportation facilities offered by buses, trucks, private autos, pipelines, and airplanes were not yet serious competitive threats to the railroads. As America faced the problems of a world at war, her railroads were still living in a golden age.

The half-century was in other ways a period of retrogression. Many, perhaps most, of the post–Civil War railroad lines suffered from the evils of inflated construction costs, fraudulent stock manipulations, and incompetent management. Such speculators as Jay Gould, Jim Fisk (1834–72), and Daniel Drew became expert jugglers of corporate property. And even more sober and respected members of the rail management group, such as James J. Hill of the Great Northern, or James C. Clarke (1824–1902) of the Illinois Central, followed patterns of business ethics that would not be tolerated in the twentieth century. Concurrently, discrimination in freight rates between individuals, localities, and articles forced merchants, farmers, and communities to agitate for government regulation of the railroads. Once people secured a measure of rail service, they tended to grow hostile toward the railroads, especially if times were bad. James C. Clarke of the Illinois Central saw this public hostility as inherent when he wrote to Stuyvesant Fish (1851–1923) in 1883: "The people are in favor of building a new road and do what they can to promote it. After it is once built and fixed then the policy of the people is usually in opposition." But a dominant theme in the last decades of the nineteenth century was that of corruption and rate discrimination, and the increasing governmental regulation which followed was a natural result.

The years of monopoly, corruption, discrimination, and ever increasing regulation were also the years in which the prevailing fashion in American economic philosophy shifted from a pronounced laissez faire doctrine to one of increasing governmental restraint and control. The American rail network was completed in the very years when the Progressive Movement was reaching its climax. It is certainly unfair to judge the business ethics of the rail management of the sixties, seventies, and eighties solely by a set of rules and values which became standard and were accepted only in a later generation. Actions of "smart" operators which were honored and emulated in the early years after the Civil War would have resulted in legal prosecution if pursued a generation later. Nor should it be forgotten that the very impatience of the public for an immediate and complete rail service in the years after the Civil War aggravated a situation in which rail management was hindered by a lack of adequate experience in both building and operating a facility of such size.

There was corruption in railroad management throughout the nation in the years after the Civil War, but some of the earliest and some of the worst accompanied Reconstruction in the eleven Confederate states. Ruined as they were by the war, southern railroads put most of their energy into the restoration of old lines in the decade after 1865. New construction in the southern states was extremely slow compared to that in the rest of the nation. In the first postwar decade, although the national rail mileage more than doubled, southern mileage increased from 9,071 miles to 14,421 miles, a gain of only 59 per cent. Illinois alone laid as much new track in the decade as all of the Confederate states east of the Mississippi. Moreover, the late 1860s and early 1870s were years of intense and profitable railroad activity by carpetbaggers and their white and African-American resident collaborators. The visitors from the North built little new mileage, for they were more interested in raiding a state treasury or milking a railroad exchequer than they were in the hard work of actually building a railroad. Many southern states paid out hundreds of thousands of dollars for railroads that were never completed or even honestly started.

Most of the carpetbag activity in the railroad industry was confined to the six coastal and eastern states running from Virginia to Alabama, but the experiences of North Carolina and Georgia are typical. In the years 1868–69 the legislature of North Carolina authorized the issuance of $27,850,000 in state bonds to thirteen different railroads in the state. Of those authorized, bonds in the amount of $17,640,000 were actually issued to railroads, but only 93 miles of line were built in the years 1869 through 1871. Instead, most of the proceeds from the sale of the bonds went into the pockets of a railroad ring, where it was spent for such things as gambling junkets in New York City and bribery in North Carolina. The suave and genial General Milton S. Littlefield (1830–99) of Illinois and New York was the active leader of the ring. Control of the pliable legislature of the state was made easier by the governorship of W. W. Holden (1818–92), a personally honest man who was generally weak and incompetent. While General Littlefield dispensed free wine, liquor, cigars, and good fellowship from a bar set up in the west wing of the Statehouse, George Swepson (1819–83), paymaster of the ring, was paying out over $133,000 in more direct gratuities to a dozen or more legislators and officials. Naturally, Littlefield and Swepson did not receive all the state railroad aid, but they were issued $6,367,000 in crisp new state bonds for their chosen road, the Western North Carolina. The bonds were sold for about 53 cents on the dollar. Very little of the more than $3,000,000 was used to give western North Carolina more railroad.

Things were no better in Georgia. The state did add substantial rail mileage during Reconstruction, but it was in spite of the carpetbagger rather than because of him. Prime mover in the corrupt railroad activities was Hannibal I. Kimball (1832–95), one-time carriage-maker in Connecticut. After the war he shifted to selling railroad sleeping cars but sometimes sold the cars without bothering to complete delivery. In Georgia between 1868 and 1871 he received much state railroad aid without bothering to build the railroads. The tall, handsome promoter was munificently openhanded with all the Republican officeholders who could help him. Several times he made substantial cash deposits to a bank account opened freely to the governor of the state, the ample and amiable Rufus B. Bullock (1834–1907), also a short-time resident of Georgia. Kimball's persuasive extravagance was so great that fellow citizens claimed the only way to resist him was to refuse to see him. During the Bullock-Kimball regime potential state aid of nearly $40,000,000 was authorized for thirty-seven different companies, nearly all of which had little more than a paper existence.

Corruption was just as rife on the state-owned Western and Atlantic Railroad as elsewhere. Mismanagement became colossal as old employees were dismissed to make way for political "patriots," the well-paying (by the sale of tickets en route for cash) conductor jobs being reserved for the sons of legislators. The company auditor explained to a subsequent investigating committee that he had saved $30,000 in a year or two out of a $2,000 annual salary only by practicing the "most rigid economy." Public clamor against the graft became so great that by late winter in 1870 the legislature was willing to lease the road if it could be granted to the "right people." The successful syndicate included H. I. Kimball, plus the more respectable conservatives Joseph E. Brown (1821–94) and John P. King of the Georgia Railroad. Clearly, the excesses of the carpetbag and radical regimes in the South helped set a new low for the business ethics of the entire nation in the postwar generation.

In the northern and eastern states after the Civil War the corruption and chicanery found in much of the railroad management were almost beyond comparison. Where Kimball, Swepson, and Littlefield had stolen thousands of dollars or here and there a million from short railroads in single states, Vanderbilt, Drew, Gould, and others were making millions and tens of millions in manipulating rail systems extending for thousands of miles. Since the rail lines in the trunk-line region between New York City and Chicago were big systems with a heavy and profitable traffic, it was natural that the corruption, when present, was also immense. Watered stock, stockmarket rigging, corrupt and dishonest management, rate wars, rebating, and labor violence were all part of the railroad picture in the postwar generation.

The oldest and perhaps the greatest of the northern railroad barons was Commodore Cornelius Vanderbilt. At the ripe old age of sixty-eight, already possessing an $11,000,000 fortune gained from shipping and steamboats, the Commodore turned to railroads in 1862. Conquering a long-time aversion to railroads, which might be traced back to a serious injury received in a wreck on the Camden and Amboy in 1833, the old man secured control of the New York and Harlem for no more than $9.00 a share and the New York and Hudson for about $25.00 a share. The two lines provided service up the Hudson to Albany. Erastus Corning's New York Central had a habit of transferring freight and passengers to Vanderbilt's line at Albany only when the Hudson was frozen and the boat lines were unable to operate. This the Commodore did not appreciate, and in midwinter (January 1867) he suddenly stopped service between his own lines and the New York Central at the Albany bridge. Missing the leadership of Corning, who had retired because of poor health in 1864, and also that of his tough-minded successor, Dean Richmond (1804–66), the leading stockholders of the New York Central requested Vanderbilt to vote their proxies along with his own recently acquired stock in the annual election of December, 1867.

As the new president, Vanderbilt, with the expending of some money on the legislature in Albany, obtained the right to consolidate his two major roads into the New York Central and Hudson River Railroad. The Commodore ran his line with verve. He banned the use of color and brass ornaments on his locomotives and voted himself a bonus of $6,000,000 in cash and $20,000,000 of new watered stock (with perhaps $20,000,000 more for the other shareholders). The old man proceeded to make the railroad efficiently earn and pay, even in the depression seventies, annual dividends of from 6 to 8 per cent on the inflated capitalization. The Commodore was not alone in his stock-watering. In May, 1869, the *Commercial and Financial Chronicle* reported that in less than two years twenty-eight railroads had increased their capital from $287,000,000 to $400,000,000.

In his management of railroads the senior Vanderbilt was much aided by his eldest son, William H. Vanderbilt (1821–85), who returned to the inner family circle from his Staten Island exile after he had so dramatically rehabilitated the Staten Island Railroad. William soon became vice-president of both his father's first two railroads, and later, in the seventies it was at his insistence that the Commodore extended his lines to Chicago by buying the Lake Shore & Michigan Southern and the Michigan Central. All the roads prospered under the Vanderbilt management, as is clearly shown by the growth of William's inheritance of perhaps $90,000,000 in 1877 to a fortune of $200,000,000 at his own death in 1885. Father and son were remembered by the public not only for their railroad fortune and for their somber locomotives, which railroad men called "Black Crooks," but also for their language. The old Commodore's "Law! What do I care about law? Hain't I got the power?" was undoubtedly apocryphal, but son William probably did say, "The public be damned" in an interview with reporters in his private car near Chicago. William H. clearly thought the New York Central should be run for the benefit of the stockholders.

The struggle for the control of the Erie Railroad in the decade after the Civil War was one of the most spectacular chapters of nineteenth-century American finance. Once in control of the New York Central, Vanderbilt naturally was interested in the future of the Erie, his major competitor in the state. But the man who controlled the Erie was that ancient rustic who looked like a country deacon, Daniel Drew. "Uncle Daniel," former drover of cattle and "watered" stock and long-time rival of Vanderbilt in the steamboat business, invested heavily in Erie stock in 1854, the same year that Mrs. Silas Horton had prevented a rail disaster by flagging down an Erie train with a pair of her red woolen drawers. In 1867, when the Commodore was buying Erie stock, Drew was aided in his management of Erie by a pair of rascals from Vermont and Connecticut. Jim Fisk, roly-poly son of a Vermont peddler and all brass, had served an apprenticeship to his later financial buccaneering as ticket-seller for Van Amberg's circus. He had also been a tin-peddler and a buyer of Confederate cotton before becoming a stockbroker in Wall Street. Jay Gould, anything but an extrovert, was to prove a better manipulator of stocks than either Fisk or Drew. Gould and Fisk had already helped Drew unload on the Erie the small Buffalo, Bradford, and Pittsburgh Railroad. The purchase price of $2,000,000 in Erie convertible bonds was a sum eight times the true value of the acquired road.

Finding the Commodore still eager to buy Erie stock, the trio unloaded 100,000 crisp new shares (converted from Erie bonds) on the old man. Vanderbilt, feeling cheated, in early March, 1868, obtained from his good friend Judge George G. Barnard (1829–79) a warrant to arrest Gould, Fisk, and Drew. The trio ferried across the Hudson and found safety in Taylor's Hotel in Jersey City, surrounded by $6,000,000 in cash, more unsold bonds, and a large body of guards commanded by Fisk. From the safety of "Fort Taylor" the breezy Fisk gave the following statement to the press: "The Commodore owns New York, the Stock Exchange, the streets, the railroads, and most of the steamships there belong to him. As ambitious young men, we saw there was no chance for us there to expand, and so we came over here to grow up with the country."

Gould hurried to Albany, where the expending of at least $500,000 on the legislature and assistance from Senator (Boss) William M. Tweed (1823–78) legalized the whole affair. The old Commodore admitted defeat and retreated to his New York Cen-

End of track, 1867, near Archer, Wyoming. Supplies for the construction of the Union Pacific came both by rail and by covered wagon. (Courtesy, Union Pacific Railroad.)

Six of a kind. These woodburners at Waseca, Minnesota, in 1881 permitted little in the way of "featherbedding" by crew members. Even the turntable required plenty of manpower. (Courtesy, Chicago & North Western Railway.)

tral, muttering that his trouble with Erie, "has learned me it never pays to kick a skunk." Drew, now replaced in top Erie leadership by Gould, turned to less successful exploits on Wall Street, where he was finally ruined in the Panic of 1873, left with little more than his Bibles, hymnbooks, and sealskin coat. Fisk, the new vice-president, was known as "Prince of Erie" and had a marble and gilded office in Pike's Opera House, which he and Gould rented to Erie at the not too modest rate of $75,000 a month. He went on to be "Admiral" of his own steamship line and Colonel of the 9th Regiment, New York National Guard. But his flamboyant career was ended early in 1872, when he was shot by a rival for the affection of his favorite mistress, Josie Mansfield. The Erie Railroad lay in ruins, and Gould soon took his profits elsewhere.

Karl Marx once described the experienced man of business as possessing fine hearing and a thick skin, a man capable of being venturesome and cautious simultaneously, a swashbuckler and a calculator. He might well have been describing Jay Gould as that gentleman maneuvered the fortunes of the Union Pacific in the seventies. By 1874, Gould controlled this Pacific road, having bought the shares, beset as they were by the Panic of 1873 and the memory of the Credit Mobilier, for as little as $14.00 a share. He then forced the Union Pacific to take at par the shares of both the Kansas Pacific and the Denver Pacific, roads he had purchased for a few dollars a share early in the depression. It had been looted long before, but Gould contrived to extract from the Union Pacific more than $10,000,000 before the management was taken over by Charles F. Adams, Jr., in 1884. As Drew had said of Gould, "his touch is death." In the eighties, Gould, whose frail body was already worn thin by advanced tuberculosis, was just as ruthless in his control of the Wabash, the Missouri Pacific, and the Texas and Pacific. When he died in 1892 at the age of fifty-six, Gould, who by this time was being called a pirate and a scourge by the *New York Times*, left a fortune conservatively estimated at $77,000,000.

Commodore Cornelius Vanderbilt. The Commodore's first real interest in railroading came in 1862, when he was sixty-eight and already a multimillionaire. (Courtesy, New York Central System.)

Jay Gould. In the years after the Civil War this crafty stock manipulator controlled in turn the Erie, the Union Pacific, and the Missouri Pacific. (Courtesy, Erie Railroad.)

Thomas A. Scott. Like
Gould, Scott had interests
in many railroads, but his
major concern was with
the Pennsylvania.
(Courtesy, Pennsylvania
Railroad.)

Edward H. Harriman.
In the early twentieth
century Harriman
controlled a network of
25,000 miles which
included the Union
Pacific, the Southern
Pacific, and the Illinois
Central. (Courtesy, Union
Pacific Railroad.)

There were other railroad abuses in the postwar years besides the excessive watered-stock operations of Vanderbilt, the thimble-rigging market operations of Drew, and the general chicanery of Gould. In the sixties, seventies, and eighties many railroads participated in rate wars, which resulted in a general disturbance of normal business. During such "wars" the complaint of shippers and merchants was not so much against the published rates as against the fluctuation and uncertain departure from established rates. Shippers in the New York City–to–Chicago area complained that the published rates often varied from forty to sixty times in a single year.

Naturally, the bulk of the rate-cutting occurred in competitive regions where several lines served such routes as Chicago to New York City, Chicago to St. Louis, or coast to coast. When the Erie and New York Central were fighting, the Commodore finally slashed his livestock rates between Buffalo and New York City to $1.00 a head. The canny Jim Fisk retaliated by making heavy cattle purchases in Buffalo, after which he very profitably shipped his livestock to market over his rival's road. Passenger fares were also cut. A ticket from New York to Chicago reached a low of $5.00 in 1881, and at one time in 1884 the Chicago–to–St. Louis fare dropped to $1.00. The transcontinental traffic had been so heavy in the spring of 1886, when a New York–to–California ticket cost only $29.50, that a western rail official claimed that they were now carrying "tramps."

Many of the rate wars were aggravated in total effect by the practice of bankrupt lines, or those in receivership, charging very low rates. Temporarily free of meeting their interest payments, such lines could afford to reduce their rates, hoping for preferred treatment when rate stabilization occurred or the next railroad pool was arranged. In 1882, Vanderbilt bought up such a line competing with his Lake Shore road when he purchased control of the Nickel Plate (New York, Chicago, and St. Louis) from a syndicate headed by George I. Seney (1826–93) and Calvin S. Brice (1845–98). Seney and Brice, who were becoming known for their fast-and-loose management of southern railroad property, were indifferent about whether they sold their road to Gould or Vanderbilt. When Vanderbilt complained that the Nickel Plate was bankrupt, Brice clinched the deal by answering: "No one knows that better than I do, but do you want to compete with a receiver?"

An obvious answer to rate wars was the railroad pool. The public naturally objected to pooling because it tended to keep rates high, but the typical railroad executive favored the pool as the best solution to cutthroat competition. One of the earliest pools was established among three western lines, the Burlington, the North Western, and the Rock Island. Having approximately equal facilities for handling the through traffic between Chicago and Omaha, the three roads decided in 1870 that each should have a third of the business. This arrangement, known as the Iowa Pool, was maintained with little interruption until it gave way to a larger organization, the Western Freight Association, in 1884. In the southern states rate stabilization was achieved with the organization in 1875 of the Southern Railway and Steamship Association. Colonel Albert Fink (1827–97), former general superintendent of the Louisville & Nashville, was so successful as the first commissioner that twenty-seven railroads were members by 1877. Colonel Fink soon moved on to administer a still larger pool covering the trunk lines between New York City and Chicago. But even Fink could not assure a permanent peace among the eastern lines.

Some of the rate wars occurred as rival systems built, or threatened to build, "nuisance" or parallel lines in a competitor's area. In the early eighties such a struggle developed between William H. Vanderbilt's New York Central and the Pennsylvania Railroad. In 1881 the New York, West Shore, and Buffalo was chartered to build a line on the west side of the Hudson, paralleling Vanderbilt's property all the way to Buffalo. The West Shore, originally headed by General Horace Porter (1837–1921), was partially financed by George M. Pullman, who was angry because Vanderbilt had used his own Wagner sleeping cars instead of Pullmans on several New York Central subsidiary lines.

The Pennsylvania soon acquired a financial interest in the West Shore. The line was no stranger to such projects. Back in the seventies, when Tom Scott had dreamed of a Pennsylvania-dominated southern rail empire, he had built, or threatened to build, parallel lines in Maryland and Virginia against John W. Garrett's (1820–84) Baltimore & Ohio and William Mahone's (1826–95) Atlantic, Mississippi, and Ohio. When Vanderbilt realized who his major antagonist was, he retaliated by going into Pennsylvania. With financial aid from Andrew Carnegie (1837–1919), who was unhappy with the existing rail monopoly of the Pennsylvania in Pittsburgh, Vanderbilt started to build the South Pennsylvania Railroad just a few miles south of the main line of the Pennsylvania. The rate war and struggle between Vanderbilt and the Pennsylvania grew so bitter that the New York Central reduced its dividend, and the country's top banker, John Pierpont Morgan (1837–1913), decided to intervene. Morgan had gained the confidence of Vanderbilt several years earlier when he had successfully sold a large block of New York Central stock in Europe. In July, 1885, Morgan invited Vanderbilt, Chauncey Depew (1834–1928), now president of the New York Central, and George B. Roberts (1833–97), Scott's successor as the Pennsylvania president, to a conference on his palatial yacht the "Corsair." While the boat slowly cruised in the East River and the Sound, an agreement was finally reached. The nearly bankrupt West Shore was taken over by the New York Central, while the Vanderbilt line abandoned its partly completed road in the Keystone state to the Pennsylvania. Morgan received fees estimated at from $1,000,000 to $3,000,000 for running the peace conference and carrying through the necessary subsequent corporate reorganizations. He also emerged as a new force in American railroading.

Corruption, Discrimination, and Regulation

One of the most obnoxious types of rate discrimination was the rebate. Rebates had been granted occasionally before the Civil War, as in the case of several anthracite lines in Pennsylvania in the middle fifties. It remained for John D. Rockefeller (1839–1937) and his Standard Oil Company to make it really effective. Rockefeller's partner, Henry Morrison Flagler (1830–1913), started to bargain for freight rebates in 1867. Flagler was so successful that his company was soon receiving not only rebates, or refunds on its own freight costs, but also drawbacks, or refunds on the higher freight paid by competitors. In Ohio, for example, while Standard was paying 10 cents a barrel, competitors paid 35 cents a barrel, with the difference, 25 cents, being refunded to Rockefeller's company. In a year and a half, four railroad companies paid Standard over $10,000,000 in drawbacks. Rockefeller's oil monopoly often forced railroads to allow it the exclusive use of tank cars at a lower rate, while independent companies were relegated to wooden barrels at a higher rate.

In the years after the Civil War most of the other segments of American industry were limited in their acquisition of rebates and special rates only by their own economic leverage. William H. Vanderbilt admitted to the Hepburn Committee in New York in 1879 that all large shippers who applied for special rates normally received them. In the first six months of 1880 his company granted six thousand special rates.

A public opinion already critical of railroad management was sometimes further incensed by occasional labor disputes, especially those of 1877 and 1894. Railroad labor was far from fully organized in the early postwar years. The Big Four Brotherhoods (Engineers organized in 1863, Conductors in 1868, Firemen and Enginemen in 1873, and Trainmen in 1883) had been established mainly as mutual-insurance societies, since the risks of railroad employment made commercial insurance expensive. The trouble in 1877 started after most eastern railroads had reduced wages.

The Baltimore & Ohio in July, 1877, announced its second 10 per cent wage cut and also started to double the length of its freight trains without increasing the size of the crew. The firemen and brakemen on the B.&O. struck. The violence along John Garrett's line was nothing compared to the mad fury of the Pittsburgh mobs which besieged the militia in a roundhouse and eventually destroyed more than $5,000,000 worth of property.

The trouble spread to Buffalo, Chicago, St. Louis, Omaha, and St. Paul. Militia, policemen, regular-army regiments, and even GAR veterans were used to stop the violence, 10,000 of them being required to reopen the main line of the Pennsylvania. The strike was broken, but the railroad workers and much of the general public could not help recalling the huge estate of the late Commodore and the dividends still being regularly paid by the B.&O., the New York Central, and the Pennsylvania. Another labor dispute of great intensity occurred in 1894 when thousands of rail workers halted hundreds of trains in sympathy with the Pullman strike. After peace had been restored, many of the strikers, members of Eugene V. Debs's (1855–1926) American Railway Union (150,000 members), were "blacklisted" from future rail employment when their service letters were written on stationery carrying a secret watermark showing a swan with a "broken neck."

The western farmer in the generation after the Civil War read about the Erie War, Vanderbilt's watered stock, stockmarket operations, the rate wars, and the rebates given Standard Oil. He knew these things adversely affected him, and he also knew that he had a railroad problem much closer to home. Most of the eastern rail abuses were also found in western railroads. In addition, the western farmer's problem was aggravated by other factors. The Nebraska or Iowa farmer in the seventies faced a different situation from that which his father or grandfather had faced half a century earlier in a more easterly farming frontier. When inland transportation facilities, i.e., the canal or turnpike, were available to the early nineteenth-century farmer, the employment of a commercial carrier was a convenience, not a necessity. If he had to, this earlier farmer could, with his own wagon or barge, move his crop to market. Not so with the Iowa or Nebraska farmer. No longer a self-sufficient producer but, rather, a man producing for a distant market, he had no choice but to use the rail facilities offered at the rates ordained by a largely absentee ownership. It was the rare western farmer who had the choice of two rail routes to market. The added competitive facilities of lake, canal, or river barge, present east of the Mississippi, were totally lacking on the western plains. Furthermore, the greater distances, the relatively sparse population, and the absence of any significant industrial rail traffic all resulted in a level of average western rail rates substantially higher than those typical in the East. As Frederic Logan Paxson has pointed out, for the western farmer in the last part of the nineteenth century, transportation was a tax, a totally unavoidable one.

The ultimate anger of the western farmer with the railroads was the more bitter because his need for them was so great and his original expectations had been so high. In the fifties and sixties, when he had few railroads, the farmer was so hopeful of achieving cheap transportation that he was not only willing to permit his town and county governments to help finance them, but he also frequently mortgaged his farm to buy railroad stock. In the early rail-building in Wisconsin, 1,300 farmers in eight counties mortgaged their farms for a total of more than $1,500,000 to purchase such stock. Around Watertown, Wisconsin, they mortgaged their farms for stock in a railroad that never materialized. Even when the lines were built the farmer often found his land mortgaged, his railroad stock of little value because of excessive "water" or a reorganization, his taxes high because his township had also helped the railroad, and his transportation costs still excessive.

The farmer's transition from railroad proponent to railroad antagonist came earlier along the eastern prairie than it did farther west. The shift came in Illinois by 1869 or 1870, while in Nebraska, which still hopefully wanted more railroads in the early seventies, the hostility did not become intense until later in the decade. In all states the bad times that followed the Panic of 1873 caused an increase in the criticism of railroads and their rates.

Farmers blamed low farm-produce prices on high freight rates. They told each other about the Illinois farmer who returned from town with a pair of shoes for his boy as the only purchase possible from a load of grain he had sold or the Iowa farm family who burned corn for fuel because at 15 cents a bushel it was cheaper than coal. The farmer might thus be warm, but he got hotter when he reflected that corn was selling for $1.00 a bushel in the East. The rates were high at non-competitive points (and this was the typical small farm town) because of long-and-short-haul discrimination. The grain rate from the Twin Cities to Chicago in the nineties was 12.5 cents a hundred pounds, while the rate from many Minnesota towns to St. Paul was 25 cents a hundred. And Wisconsin farmers on any one of the three competing lines into Chicago also paid a higher local rate for a shorter haul. In 1869 roads from the Mississippi to Chicago charged 20 cents a bushel for a 200-mile haul, while grain could be carried by rail to the seaboard for twice as much. Rates for the shipment of cattle in Illinois were so high that farmers frequently drove their stock to the Chicago market.

To compensate for low competitive long-haul rates and the frequent rate wars, the railroads were pushing up their local rates to any level that would not completely choke off business. Their rule for making rates was: "Charge all the traffic will bear." The complaint against the long-and-short-haul discrimination was not a complete monopoly by the western farmers. Farmers at non-competitive points in New York state felt that they were farther from market than some of their well-located western competitors. A tub of butter could be sent from Elgin, Illinois, to New York

City for only 30 cents, while the same tub from some points no more than 165 miles out of New York City cost 75 cents to send to market.

The farmer had other complaints. He hated to be at the mercy of the local elevator company, often owned by the railroad and managed by a man who downgraded the farmer's grain and who could frequently afford to hold it until a profitable price rise came along. Nor did the farmer like the fancy prices he paid at the local country store, prices high in part because the merchant was in the same freight-rate bind as the farmer.

Farmer and local merchant alike, in fact all the little people, were unhappy about the abuse of railroad passes. All sections of the country complained about the influence railroads exercised over public officials through the universal practice of granting free passes to congressmen, judges, sheriffs, assessors, and even town officials. The railroads were just as generous with members of the press, and they also favored big shippers and agents of large concerns. In the generation after the Civil War this abuse grew rapidly. By 1897 the railroads of North Carolina were giving out 100,000 passes a year at an estimated annual revenue loss of $325,000. Paying passengers found it difficult to conceal their disgust when they found their seatmate to be a "deadhead," for they knew they were indirectly paying his passage. Some railroad officials were well aware of the public resentment. James M. Walker (1820–81), president of the Burlington in the early seventies, wrote his vice-president: "I think the grant of a free pass to make one friend creates half a dozen enemies." But the friends were frequently very valuable. Railroads saved thousands of dollars in taxes with passes given county assessors and auditors. As the custom became intrenched, public officials came to view the annual pass as a vested right. This railroad evil of mixed blackmail and bribery remained to plague the public until it was finally prohibited by the Hepburn Act of 1906.

The main source of railroad abuses and difficulties in the late nineteenth century was their freight rates. The inherent complexity of the subject contributed to the difficulty. An author reviewing the problem early in the twentieth century claimed that there were about fifteen thousand principles governing the movement of interstate railroad freight traffic. This was perhaps an exaggeration, but, clearly, as the carrier, the shipper, and the general public sought a reasonable level of railway charges for a traffic of unlimited variety moving over dozens of railroads between hundreds of communities and trading centers, any thought of a simple solution was pure folly. By 1900 a great variety of principles or theories of rate-making had appeared. Such rate theories as cost of the service, flat rate by distance, flat rate regardless of distance, what the traffic will bear, keep everybody in business, value of the service, value of the commodity, and the general public interest were advanced by various interested parties. Of the several theories, the railroads favored charging as much as the traffic would permit. At the turn of the century Charles A. Prouty (1853–1921), member of the Interstate Commerce Commission, asked a railroad traffic official the basis on which his rates were made. The official replied: "To be perfectly honest, we get all we can, and even that is too little." A more reasonable and equitable rate would be one which considered the interests of the shipper, the railroad, and the general public. Placed somewhere between the minimum of the out-of-pocket costs to the railroad and the maximum of the total value of the service to the shipper, the rate would finally be determined by the conditions of competition, the value of the commodity, and the importance of its movement to society.

Such ideals in the making of rates were obviously not present in the generation after the Civil War. Rate discrimination took three forms in the period: between classes of traffic or freight, between places, and between persons. There was nothing bad in the principle of charging more for some commodities than others. Dry goods or general merchandise could obviously absorb a higher rate by weight than such goods as lumber, coal, or grain. There was also some justification for some of the place discrimination. Clearly writing from the railroad point of view, General Edward P. Alexander (1835–1910), president of the Georgia Railroad, answered critics of the long-and-short-haul discrimination in the *Commercial and Financial Chronicle* in 1887: "None of these discriminations made between places are, however, the result of a wish on the part of the railroad to discriminate, but arise from the fact that nature has discriminated by giving the river town natural transportation." Much of the local or place discrimination in the period, whether due to "nature" or not, was excessive and unjustified. Except for lower rates given for quantity shipments, there was no justification at all for the many cases of rate discrimination between persons. Where special rates were given, they were generally secret, often marked by rank favoritism, and often accentuated inequalities, given as they were to those who needed them least.

Before the Civil War several eastern states had supervised their railroads, but never strictly. Railroad commissions or regulatory bodies were established in Rhode Island in 1839, New Hampshire in 1844, Connecticut in 1853, Vermont in 1855, and Maine in 1858. In general the supervision was established to check on the compliance with the railroad charters, to provide for just apportionment of interstate receipts, and to reduce the incidence of railroad accidents. Another aspect of railroad operation, that of Sunday trains, while perhaps rarely the concern of state agencies, was a heated issue in many localities. In Vermont in the fifties the legislature decided that conductors were to read a passage of Scripture to passengers traveling on Sunday trains. Erastus Fairbanks (1792–1864), scale manufacturer, one-time governor of Vermont, and a strict observer of the Sabbath, made certain that no trains ran on Sunday on his road, the St. Johnsbury and Lake Champlain. Keepers of the Sabbath also stopped train service on Sunday in Altoona,

Pennsylvania. In Illinois the president of Knox College was less successful when he attempted to stop Sunday train operations through Galesburg. The railroads started to give the clergy passes, the Civil War years saw little agitation against Sunday trains, and after the war the railroads tended to run trains how and when they pleased.

After the Civil War, complaints against more serious aspects of the railroad problem led to a growing demand for regulation. President Grant urged Congress in 1872 to investigate the problem of cheaper freight rates for western and southern products. The resulting Senate committee, headed by William Windom (1827–91) of Minnesota, in its 1874 report criticized the railroads for several abuses but specifically recommended little more than the construction of several new water routes to the Atlantic coast. In 1879 the New York legislature decided to investigate the railroad problem, and a committee under the chairmanship of Alonzo B. Hepburn (1846–1922) obtained much information, especially on the evils of rebates and rate discrimination. In spite of the public outcry at the revelations in these reports, the eastern states seemed satisfied to depend upon publicity and the force of public opinion. Most railroad regulation by eastern states in the seventies and eighties followed the pattern set by the Massachusetts commission of 1869, an agency largely advisory in nature. It remained for the western and southern states to take more positive action in railroad regulation.

The first significant regulation of railroads after the Civil War appeared in the seventies in several prairie farming states. The railroad laws that were enacted in the upper Mississippi Valley in these years are often called "Granger laws" because they were possible only through the active support of members of the Grange. The National Grange of the Patrons of Husbandry was founded in 1867 by Oliver H. Kelley (1826–1913), a Minnesota farmer who had become a clerk in the Bureau of Agriculture in Washington,

D.C. Kelley resigned his Washington post in 1868 to devote his full time to the establishment of local Granges. The work was slow and discouraging, but by the end of 1869 he had set up thirty-seven local Granges in his own state of Minnesota and the next year had units in eight other states. At first the organization was intended to be only social and educational in nature, but the hard times of the seventies expanded not only the size but also the objectives of the group.

By 1875, when the Grange had blossomed to 20,000 local Granges and 800,000 members, its attention had clearly shifted to economic and political issues. As Grange members listened to home-grown orators recite the oppressions farmers suffered at the hands of the "non-producers," their eyes probably shifted to a brightly colored lithograph on the wall of their dimly lit hall. The poster showed eight men in characteristic poses: "I pray for all," said the parson; "I trade for all," said the merchant; "I plead for all," said the lawyer; "I legislate for all," said the statesman; "I prescribe for all," said the physician; "I fight for all," said the soldier; "I carry for all," said the railroad owner; and in the center stood the farmer, who said, "And I PAY for all." By the seventies the farmer was tired of paying.

Stringent railroad regulation at the state level began in Illinois with legislation in 1871 and 1873. The legislation set up a board of railroad and warehouse commissioners, established maximum passenger fares, provided that freight rates should be based entirely on distance, and finally, in 1873, made it the duty of the new commissioners to prepare a schedule of reasonable maximum freight charges. Farmer and Granger political activity was important in obtaining this railroad legislation, the number of local Granges in Illinois in 1873 being exceeded only by the number in Iowa. Minnesota also passed legislation in 1871 fixing railroad rates and providing for a railroad commissioner. Finding its railroad laws difficult to enforce, in 1874 the state legislature modeled their regulation on the Illinois law of 1873. Regulation in Minnesota was more difficult than in Illinois because no single set of rates could equitably be applied in a state which varied from regions of heavy population to areas that were still mainly frontier.

In 1874 Granger railroad legislation was enacted in both Iowa and Wisconsin. When Iowa had earlier turned the land grants obtained from the federal government over to her new railroads, it had been with the reservation that the companies would be subject to subsequent legislative action. The Iowa Acts of 1874 were certainly satisfactory to the farm group, since the majority of the members of the lower house were Grangers. The legislation was carefully drawn and appears to have been generally enforced, since the railroad managers admitted that the state's position was invulnerable. In Wisconsin the Potter Law of 1874 established many railroad freight rates that were unreasonably low. On the advice of eminent legal counsel, Alexander Mitchell, president of the Chicago, Milwaukee & St. Paul, and Albert Keep (1826–1907), president of the Chicago & North Western, wrote the governor that they intended to ignore the legislation. The state forced the roads to comply with the law, but in 1876 the law was modified to the general satisfaction of the railroad interests. In the late seventies or early eighties Granger agitation for railroad regulation was also successful in such neighboring prairie states as Nebraska, Kansas, and Missouri. In 1879 Georgia set up a railroad commission to establish rates, and in the same year a new constitution for California provided that an elective commission should fix maximum railroad rates. Western farmers and the Grange had brought regulation to the railroads.

Western railroads vigorously resisted the enforcement of the Granger laws in a variety of ways. In the courts they argued that since their business was a private one the states had no power to fix rates. The railroad lawyers claimed that the states, which had granted railroad charters (which included the right to set rates and fares), could not now fix rates themselves without being guilty of violating the original contract. They cited the Dartmouth College case of 1819 in support of their contention. In the contest, the Grangers had votes; the railroads possessed money. While there were undoubtedly more charges of bribery than actual cases, many railroad favors were received by willing legislators. In Iowa passes were refused those politicians who were no longer friends of the railroads. In Minnesota and Wisconsin the railroads made their compliance with the new laws as obnoxious as possible. In its zeal to have uniform passenger fares, the St. Paul and Pacific raised its rates between Minneapolis and St. Paul from three to five cents. In Wisconsin the railroads obeyed the Potter Law by giving the public the poorer service of "Potter cars, Potter rails, and Potter time." Railroad methods were successful, and soon not only was the Potter Law repealed in Wisconsin, but regulations were softened in several other states as well.

Even as the railroads were winning modest legislative victories, the United States Supreme Court, in a series of decisions in 1876, decided the basic issue of laissez faire versus public regulation in favor of the Grangers. In the first of the Granger cases, *Munn* v. *Illinois*, the Court upheld the Illinois law of 1871, which fixed maximum rates for the storage of grain. At the same time, the Court decided in *Peik* v. *Chicago & North Western R.R.* and *Chicago, Burlington & Quincy R.R.* v. *Iowa* that not only could states fix maximum railroad freight and passenger rates but in the absence of national legislation this regulation could apply to interstate commerce. Only in the dissenting opinion of Justice Stephen J. Field (1816–99) in the case of *Munn* v. *Illinois* was it argued that the act in question was invalid because it amounted to a deprivation of property without "due process of law." In principle at least, the Granger victory for the state regulation of railroads seemed complete.

In the decade of the eighties railroad regulation moved from the state to the national scene. A bill seeking the creation of a federal railroad commission had been introduced in Congress as early as 1871, and John H. Reagan (1818–1905), former Confederate postmaster general from Texas, in 1878 had managed to get his regulation measure through the House. In 1880, James B. Weaver (1833–1912) complained in Congress that railroads in Iowa were succeeding in shifting litigation from state to federal courts by pointing to their incorporation in another state. Voices were being raised that railroad regulation was too big a problem for the states. The railroads themselves were expanding greatly, both in mileage and in concentration of control, in the decade. The depression of 1884–85 speeded up this trend toward consolidation. Between 1880 and 1888 some 425 different corporations, or nearly a quarter of all railroad companies, by lease, purchase, or merger, came under the control of other lines. Noting the growing railroad problem, President Chester A. Arthur, in his annual message in 1883, urged that Congress act to protect the public against railroad abuses.

In 1885 a special Senate committee headed by Shelby M. Cullom (1829–1914), former governor of Illinois, conducted an investigation of railroad practices. The final report of the Cullom Committee in January, 1886, listed all the familiar abuses of unreasonably high local rates; discriminations between persons, places, and types of freight; special secret rebates and drawbacks; passes; watered stock, causing excessive capitalization; and managements that were extravagant and wasteful. The report recommended the creation of an independent commission to regulate the nation's railroads.

A second event in 1886 also indicated that federal action was near. In October in the so-called Wabash case (*Wabash, St. Louis and Pacific Railway* v. *Illinois*) the Supreme Court decided that a state could not regulate any rates on shipments passing beyond its own borders. The regulation of interstate commerce was to be left to the federal government. The fresh evidence of railroad abuses, plus the blow at state regulation, made federal action inevitable. The Cullom bill in the Senate and the Reagan bill in the House were molded by conferences into the Interstate Commerce Act, which was signed by President Cleveland on February 4, 1887.

The Interstate Commerce Act, in language perhaps deliberately vague, required that all interstate rates be "reasonable and just" and prohibited the familiar competitive practices of rebates, draw-backs, and pools. It required the railroads to publish their rate schedules, a practice rarely observed, though normally required in most of their charters, and directed the roads not only to post their schedules "in every depot or station" but also to file them with the government. Higher charges for non-competitive short hauls than for competitive long hauls were also prohibited. A five-man Inter-state Commerce Commission was created to administer the Act and enforce its prohibitions. The Commission could hear com-plaints from shippers, examine witnesses, and look into the records and books of railroad companies. It could also demand annual re-ports from the railroads and could insist, after a reasonable period for adjustment, upon the use of a uniform system of railroad ac-counting. While the Commission could not fix railroad rates, it could issue a cease-and-desist order for the carrier to reduce its charges. If the railroad refused to comply, the Commission could only enforce obedience through the federal courts.

The public reception of the railroad regulation in 1887 was generally favorable. The *Commercial and Financial Chronicle* called the legislation "paternalism in an extreme form," but the *Chicago Tribune* held that the railroad pool managers were getting only what they deserved. Many railroad men were not too perturbed as the Act neared passage. While the final conferences were taking place, James C. Clarke, president of the Illinois Central, wrote Stuyvesant Fish: "The only thing left for Rail Roads to do is to largely increase the long haul or through rates. Act honestly with each other and be patient." Railroad stocks even advanced upon passage of the Act.

Much criticism of the Act was avoided when President Cleveland promptly appointed five capable men to the Commission. By the end of March he had selected Thomas M. Cooley (1824–98), a Republican judge from Michigan and one-time railroad receiver, to head the Commission; the other members were William R. Morrison (1825–1909) of Illinois, Augustus Schoonmaker (1828–94) of New York, Aldace F. Walker (1842–1901) of Vermont, and Walter L. Bragg (1838–91) of Alabama. The *Commercial and Financial Chronicle* was happy to note that all the appointees were honest men with legal training and that three of them were familiar with railroad problems.

The "era of good feelings" toward the new Interstate Commerce Act and the Commission which was to enforce it was of short duration. Qualified though the new Commissioners were, the task of applying the law to the complicated railroad rate structure was like "cutting a path through a jungle." Just what were "reasonable and just" railroad rates? When shippers complained about a given rate, the necessary examination of connecting customs and issues might require a long series of readjustments, which would make a final decision anything but prompt. Naturally, the shipping public complained of the long delays in the proceedings. The Commission frequently found that witnesses were reluctant to testify or that shippers were unwilling to complain about discrimination. The major flaw in the total regulation lay in the lack of any effective means for the Commission to enforce the decisions it might make. Its cease-and-desist orders were often ignored by the carriers, and any resulting court action was slowed by appeals or demands for new hearings. In the nineties the average court case lasted for four years, and many dragged on much longer.

In 1896 two decisions of the Supreme Court appeared to be

giving the green light to the Commission. In the first case the Court upheld an act of Congress giving the Commission the power to force railroads to testify before it (*Brown* v. *Walker*) and in the second held out the hope of faster future proceedings by admitting that the courts should pass only upon the procedure and not the substance of each case (Social Circle case). But the great majority of railroad rate cases favored the railroads. Of the sixteen cases decided by the Supreme Court between 1887 and 1905, fifteen were in favor of the carrier and against the Commission. Decisions in 1897 and 1901 conclusively held that the Commission had no rate-making power and also that it possessed no power to prohibit the discrimination in which more was charged for a short haul than for a long one.

Clearly, the experience of the Interstate Commerce Commission in the first fifteen years of its existence justified the prediction of Richard S. Olney (1835–1917), corporation lawyer and later Attorney General under Cleveland. In 1892, Olney wrote a railroad friend advising him not to ask for the repeal of the act: "It satisfies the popular clamor for a government supervision of the railroads, at the same time that such supervision is almost entirely nominal." Nominal as this first major break with laissez faire was, it was to be the foundation for later, more effective regulation.

The more stringent regulation that came in the twentieth century was the result of two forces: first, the fresh evidence of increasing railroad corporate power accompanied by continuing abuses; and second, the appearance on the American political scene of the Progressive Movement. In the nineties and the early twentieth century the Sherman Antitrust Act of 1890 was no more successful in curbing railroad consolidation than the Interstate Commerce Act had been in establishing an effective regulation. What had been in the early nineties a large number of independent systems was brought in a decade and a half into six or eight major combinations. The severe Panic of 1893 pushed many roads into default, and by mid-1894 a quarter of the nation's railroads, representing over 40,000 miles and $2,500,000,000 of capital, were in the hands of receivers. This started the process of combination, but consolidation went into high gear later in the decade. In little over a year—July, 1899, to November, 1900—some 25,000 miles of line shifted to the control of the expanding combinations.

By 1906 nearly two-thirds of the nation's rail mileage of 225,000 miles was in the control of seven groups: (1) the Vanderbilt roads (more than 22,500 miles), including the New York Central system and the Chicago & North Western; (2) the Pennsylvania group (20,000 miles), including the B.&O. and the Chesapeake & Ohio in addition to the parent line; (3) the Morgan roads (18,000 miles), controlling the Erie, the newly organized Southern, and half a dozen other southern lines; (4) the Gould roads (17,000 miles), managed by George J. Gould (1864–1923), son of Jay, and including the Missouri Pacific and other southwestern lines; (5) the Rock Island system (15,000 miles), put together by William H. Moore (1848–1923) and important in the Mississippi Valley; (6) the Hill roads (21,000 miles), including the Great Northern, Northern Pacific, and Burlington; and (7) the Harriman lines (25,000 miles), consisting mainly of the Union Pacific, the Southern Pacific, and the Illinois Central. Much of this concentrated management was under banker control, since such houses as J. P. Morgan and Company and Kuhn, Loeb and Company had invested heavily in the defaulting roads after 1893. Morgan had played a major role in reorganizing the Southern, the Erie, the Northern Pacific, and the B.&O. after the panic. He also had become the banker for the Hill roads of the Northwest.

Much of the new concentration of railroad control had been accompanied by financial manipulations so unscrupulous that they would have excited the envy of the elder Gould, Drew, or Commodore Vanderbilt. Edward H. Harriman (1848–1909), George J. Gould, and James Stillman (1850–1918), head of the National City

Bank of New York, were top men in a syndicate which financially ravished the Chicago and Alton between 1898 and 1905. The once well-run road was ruined as its capital structure was expanded from $34,000,000 to $114,000,000, with the insiders taking a private profit of at least $23,000,000. William H. Moore (1848–1923), professional promoter of corporations from Chicago, was just as ruthless when his syndicate looted the Rock Island in the same years. The Moore group managed with $5,000,000 of borrowed money and the clever exchange of securities between holding companies ultimately to control a rail system whose aggregate capital structure at par stood at $1,500,000,000.

While the elder Morgan was normally a sober manager of rail properties, the financial ruin he brought to the New York, New Haven & Hartford between 1903 and 1913 was worthy of a Gould or a Moore. The ICC report to the Senate in 1914 spoke of the "loose, extravagant and improvident" management of Morgan, condemned his "financial legerdemain," and criticized the "extensive use of a paid lobby" and the "profligate" use of free passes. Clearly, the financial mismanagement of American railroads in the early twentieth century deserved additional attention.

The turn of the century saw continued and widespread railroad rate discrimination. Especially after 1890, there was a general return by the railroads to the practice of rebating, in spite of the law. In addition to the old form of rebate, several convenient substitutes were invented. Preferred shippers received the same advantage through overbilling (the railroad's paying damages for the loss of non-existent items) or underbilling (carrying a larger shipment but paying for a smaller one). Free storage of a shipper's goods or elevator rebates and allowances were also used in lieu of the straight rebate. The denial of fair and equal facilities, such as the refusal to furnish cars to a shipper, was a way of discriminating against a shipper not in favor with the railroad.

In 1903 Governor Robert M. La Follette (1855–1925) of Wisconsin began an investigation of the railroads in his state. He discovered that between 1897 and 1903 rebates and unlawful discriminations in Wisconsin amounted to more than $7,000,000, with every major line guilty of these practices. Rebating was found to have continued even after the passage of the Elkins Act, although many carriers had shifted to the new technique of issuing "midnight tariffs," special low rates purposely printed to favor a large shipper and then revoked once the shipment had been made. Still another factor which favored a demand for a tighter railroad regulation was the definite increase in average railroad freight rates which started about 1900. The rise in rates was in great measure justified by higher operating costs, but to a hostile public it seemed to be a result of the growing rail consolidation.

The revival of railroad regulation came with the Progressive Movement and with the appearance in the White House of a new and energetic president, Theodore Roosevelt (1858–1919). Roosevelt first challenged the railroads when he had Attorney General Philander C. Knox (1853–1921) start a suit under the Sherman Act to dissolve the Northern Securities Company. This giant holding company for three major northwestern lines—the Northern Pacific, the Great Northern, and the Burlington—had been put together by Harriman and Hill after they had fought each other for the possession of the Burlington. Harriman, already a dominant figure in the Illinois Central, had acquired the dilapidated Union Pacific in 1897 as it was facing foreclosure and reorganization. Having rehabilitated the Union Pacific, Harriman next acquired a working control over the Southern Pacific in 1901 after the death of Collis P. Huntington, the last survivor of the Big Four. He next sought to gain entry into Chicago by controlling the Burlington, but at the same time, Jim Hill, backed by Morgan, reached out to add the Burlington to his two northern roads. The resulting battle in Wall Street finally ended when the antagonists joined to create the Northern Securities Company. In 1904 the Supreme Court, in outlawing the holding company, ordered the dissolution of the Northern Securities Company. However, the ownership of the stock of the three railroads changed very little and the Burlington remained in the hands of the Great Northern and the Northern Pacific.

Roosevelt was more successful in two pieces of railroad legislation passed in 1903. The Expedition Act of that year sped up court procedure for the handling of suits brought by the government under either the Sherman Act of 1890 or the earlier Interstate Commerce Act. A few days later Congress passed the Elkins Act, which attempted to strengthen the prohibition of railroad rebates. The new law also made railroads, as well as company officials, liable to prosecution.

The first major change in federal railroad regulation since 1887 came in Roosevelt's second term as President. In his message to Congress in December, 1904, and again a year later, Roosevelt called railroad legislation "a paramount issue" and asked that the Interstate Commerce Commission be empowered to establish reasonable railroad rates and that the carriers be prevented from granting rate favors or issuing free passes. The railroads seemed to be ready to yield to the growing demands for more effective regulation. In December 1905, the presidents of the Pennsylvania, the New York Central, and the Reading railroads announced that as of January 1, 1906, passes would be given only to their own employees. So great was the demand for legislation that when the Hepburn Act was approved in June 1906, only seven votes were cast against it in the House and three in the Senate. The Hepburn Act extended the powers of the Interstate Commerce Commission to other common carriers, such as express, sleeping-car, and pipeline companies, and increased the size of the Commission from five to seven members (raised to nine in 1917). It abolished the granting of passes (except to employees, charitable cases, and the clergy) and strengthened the law against rebates. The so-called commodity clause insisted that railroads separate themselves from such enterprises as company-owned coal mines. Most important was the fact that the new legislation empowered the Commission to establish "just and reasonable" maximum rates, and thus the Hepburn Act became a landmark in the development of federal railroad regulation.

More regulation came under Roosevelt's successor, William Howard Taft (1857–1930), who had never had any great love for the railroads. Before he became President he had written that they were monopolies run by eight or nine men who were "exceedingly lawless in spirit." As President, Taft was especially anxious to see that any new legislation included the creation of a new "Commerce Court" to hear appeals from ICC rate decisions. The Mann-Elkins bill, enacted in June 1910, by a combination of Democrats and insurgent Republicans, established such a court and also gave the Commission jurisdiction over telephone, telegraph, and cable companies. The legislation greatly strengthened the 1887 act regarding long-and-short-hauls by deleting the phrase "under substantially similar circumstances and conditions," words which had allowed the Supreme Court to make the original provision useless. The Mann-Elkins Act also permitted the Commission to suspend any new rates for as long as ten months, with the burden of proof as to the reasonableness of the proposed rates being placed upon the carrier.

When the new Commerce Court seemed to be favoring the railroads and working against the ICC, Congress refused to vote funds for its operation and it went out of existence in 1913. A year earlier the Panama Canal Act made it unlawful for any railroad to control or have any interest in common carriers by water operating through the Canal or elsewhere, where the carriers might be competing for traffic with the railroad. Finally, the Railroad Valuation Act of March 1, 1913, gave Robert M. La Follette a law he had wanted since 1906. The legislation required the ICC to assess the value of all railroad property. In reviewing the financial history of each line, the original cost, reproduction cost, depreciation, present value, and value of all grants and gifts received were to be considered. It was expected that ultimately the Commission could thus establish rates, not upon a basis of watered stock and inflated capitalization, but, rather, upon the true value of the operating properties of the railroad.

Much progress had been made in railroad regulation at the outbreak of World War I. In the early twentieth century, Americans swept up by the Progressive Movement were in no mood to have anything less than an effective and rather complete regulation of the nation's railroads. They remembered the stealing, the watered stock, the financial chicanery, and the manipulations of Littlefield, Vanderbilt, Gould, and Brice. All these men and their comrades in corruption had passed on by 1900, but in some instances their children and grandchildren were carrying on in the same tradition. The new regulation owed much to thousands of farmers who were once individually inarticulate in the political arena but who later became collectively powerful enough, as Grangers, to create the first stringent railroad regulation at the state level. Complicated as the whole field of railroad rates and regulations was, the several amendments to the original federal legislation of 1887 had established a comprehensive system of control by the second decade of the twentieth century. True, the total legislation did not provide much in the way of assuring an adequate rail service, nor were there yet any legal provisions looking toward real supervision of the carrier's financial structure, but the federal and state regulations were being enforced by commissions of the mandatory type which possessed full powers to stop discrimination and to enforce reasonable railroad rates.

There were certainly many honest railroad men in the half-century of expanding railroad service between the Civil War and the First World War. The typical railroad manager was probably just as honest as the average businessman in the generation after Appomattox. But it was a generation when the typical businessman felt that if he succeeded in making money in his business, raised a proper family, and contributed to his church, he had fully discharged his duties to society. Individual material success was more important than a social conscience. It is interesting to note that by the time the government forced such a conscience upon the railroads and their managers, the need for a continuing and comprehensive regulation was already beginning to lessen. In the teens new competitive overland-transportation facilities loomed large on the horizon. It was possible in 1914 to travel from Chicago to Utica, New York, by the new inter-urban facilities; in 1917 nearly five million automobiles were registered in the nation; and in 1918 the government started airmail service between New York and Washington. New forces were preparing to challenge the long-time dominance of the railroad.

6

Uniformity and Consolidation

In the half-century before World War I, American railroads experienced three parallel and concurrent developments. Railroads helped settle and civilize the western half of the United States in the decades after the Civil War. As we have just noted, the railroad also played a full role in the shift from a laissez faire to a partially regulated economy in the years between the first Granger legislation and the climax of the Progressive Movement. In the same fifty years, as they expanded from an eastern network of 35,000 miles to a full-grown national system of 254,000 miles in 1916, American railroads, in a third development, experienced a new integration and uniformity of operation. The resulting increases in efficiency brought by these changes and by a host of other technical advances permitted lower railroad rates. The use of steel rails, the adoption of standard gauge, the utilization of faster and more powerful locomotives pulling longer and heavier trains, the introduction of standard time, better brakes, and improved couplers all helped to create a truly national rail network. In the post–Civil War years railroads made a generous contribution to the expanding industrialization in America. They played a full role in the growth of urban population, the extension of regional into national markets, and the appearance of our modern interdependent economy.

Perhaps of first importance in the physical maturity achieved by American railroads was the extension of the rail network. This expanded from 35,000 miles in 1865 to 53,000 miles in 1870; 93,000 miles in 1880; 164,000 miles in 1890; 193,000 miles in 1900; 240,000 miles in 1910; and, finally, to an all-time high of 254,000 miles in 1916. New track was laid at an unprecedented rate. In the three decades before the Civil War (1830–60) only four years, in the prosperous fifties (1853, 1856, 1857, and 1858), had seen as much as 2,000 miles of new line constructed annually. In a comparable period after the war (1865–95) only four years (1865, 1866, 1875, and 1895) had less than 2,000 miles of new track laid yearly.

While the average annual construction in the half-century after the Civil War was well over 4,000 miles (or an average of more than 13 miles per working day), the rate of building varied greatly from period to period. Less than 4,000 miles of new line had been laid during the four years of the Civil War. In the eight years after the war (1865–73) the national network was doubled to a total of 70,000 miles of line, with more than 19,000 miles of road being added in the three years 1870 through 1872. The Panic of 1873 and the resulting rash of rail receiverships slowed construction in the remaining years of the decade to an average annual increase of less than 2,400 miles.

The decade of the eighties was the period of greatest railroad construction. Unmarred by any major depression, it saw 71,000 miles of new line built, a figure which was more than the construction of any two preceding decades in railroad history. A full recovery from the 1873 depression was indicated in 1880, when 6,700 miles of line were laid, but the next year (1881) saw a record 9,800 miles of construction, surpassing the previous 1871 high of 7,300 miles. The year 1882 set still another record, with 11,569 miles built in twelve months. The pace slowed in the middle eighties, but even in the low year for the decade, 1885, nearly 3,000 miles of line were added to the national rail system. A new pickup in construction occurred in 1886, and in 1887 an all-time high of 12,878 miles of track were laid. A majority of the new trackage for the decade was located in states west of the Mississippi, with construction being most active in the Southwest. Southern states (south of the Potomac and Ohio rivers and east of the Mississippi) added new mileage in the decade much more rapidly than any of the other older regions, doubling their rail network in the ten years to a total of over 32,000 miles.

Railroad construction never again reached the hectic rates of the eighties. The early nineties saw fairly rapid building, but the depression beginning in 1893 brought a decline that was reminiscent of the middle and late seventies. Less than 30,000 miles of road were added to the national network in the decade. In the first years of the twentieth century the pace of construction increased somewhat, and between 1900 and 1907 nearly 33,000 miles of new road were built, reaching a total of 230,000 miles of line. After the Panic of 1907, business in general had a rapid recovery, but the returning optimism did not extend significantly to the area of railroad-building. The rate of new construction continued to decline progressively until by 1917 the rate of building was slower than that of abandonment. At the end of 1916 the nation's railroads had reached a record high of 254,037 miles of road.

As the rail network increased eightfold between the beginning of the Civil War and World War I, the rate of increase varied greatly from region to region. As has been fully noted, the western expansion was the most rapid, capturing the headlines and the country's attention every time a celebration marked the completion of another route to the Pacific. As could be expected, the slowest increase was in New England, an area that had achieved a fairly complete rail service more than a decade before the Civil War.

New England trackage little more than doubled in the half-century. Only five states in the nation failed at least to triple their rail mileage from 1860 to 1910, and they were all in New England. Maine alone had a rate of construction that approached the national pattern. Some of the increase there was caused by the construction in the nineties of the Bangor and Aroostook Railroad, a line that connected Maine's largest and most northern county, Aroostook, with the rest of the state. By 1900 this new 320-mile line was not only serving the new paper mills that had come into the spruce forests of the state but was also helping the Maine potato invade new markets across the nation.

Perhaps the very absence of major railroad construction in New England hastened the trend toward consolidation among the many short roads. In any event, by 1900, half a dozen lines accounted for four-fifths of the region's mileage, and two systems controlled more than half. The New York, New Haven & Hartford, known as the "Consolidated" back in 1872 when Connecticut had finally permitted the 63-mile New York and New Haven and the 78-mile Hartford and New Haven to combine, had grown by 1900 to a system of over 2,000 miles. Despite its rapid growth the line still had a sound financial structure, was paying good dividends from a traffic that was mainly passenger, and had a local man, John M. Hall (1841–1905), as president. Morgan had not yet come to the New Haven. The Boston and Maine was the second major New England railroad. The B. and M. in 1900 had a system of owned and leased lines that totaled over 1,700 miles, had a major financial interest in the Maine Central (823 miles), and had a growing interest in the Fitchburg (457 miles). Lucius Tuttle (1846–1914) was president of both the B. and M. and the Maine Central from the middle nineties until 1910. At the turn of the century Tuttle's Boston and Maine had so simple and chaste a financial structure (without benefit of watered stock, second mortgages, or refunding certificates) that E. H. Harriman, upon hearing of it, was reported to have exclaimed: "Great Scott! Is there anything like that left out of doors?"

In the central trunk-line region of the nation the rate of new construction in the decades after the Civil War was somewhat more rapid than in New England. In the fifty years between 1860 and 1910 the railroads of the five Middle Atlantic states (New York, New Jersey, Pennsylvania, Delaware, and Maryland) expanded from over 6,000 miles to nearly 24,000 miles of line. To the west, in the five states of the Old Northwest, the expansion in the five decades was more than a fivefold increase to a total of nearly 45,000 miles in 1910. In the area between the eastern seaports of New York, Philadelphia, and Baltimore and the inland city of Chicago, railroad dominance was naturally held at the turn of the century by the two giant systems, the New York Central (10,000 miles) and the Pennsylvania (10,000 miles). Smaller but still important through routes were the B.&O. (3,100 miles) and the Erie (2,200 miles). Other small but important lines in the East included the Philadelphia & Reading (1,900 miles), the Lehigh Valley (1,400 miles), and the Delaware, Lackawanna & Western (950 miles). West of Chicago the major Granger lines all had substantial mileage east of the Mississippi River.

South of the Ohio and Potomac rivers railroad-building after the Civil War was more rapid than in any region except the West. In the fifty years after 1860, eight of the eleven Confederate states increased their mileage at least fivefold, and the area as a whole expanded its trackage more than sevenfold to a total of 63,000 miles in 1910. At least one of the states, Texas, whose 14,000 miles of line in 1910 ranked her an easy first in the nation, was almost more western than southern. But southern rail expansion was significant. Possessing but a meager rail network in 1861, the South had little opportunity to build much additional mileage in the next decade and a half of war and Reconstruction, but in the eighties southern states built much more rapidly than the nation as a whole.

By the turn of the century five lines accounted for a majority of the mileage south of the Ohio and Potomac rivers and east of the Mississippi. Largest of the five was the new Southern Railway (6,000 miles), organized in 1894 by J. P. Morgan out of the financial wreckage of the Richmond and West Point Terminal–Richmond and Danville system. Oldest of the five lines was the Louisville & Nashville (5,000 miles), dominated by its long-time president, the crusty but capable Milton Hannibal Smith (1836–1921), a man who hated passenger business decades before it became common for some railroads to attempt to avoid such traffic. The Atlantic Coast Line and the Seaboard Air Line were the newest of the major systems. Shortly after the Civil War, William T. Walters (1820–94), Baltimore merchant, aided by his friend, Benjamin F. Newcomer (1827–95), Baltimore banker, became interested in several coastal railroads running from Weldon, North Carolina, southward to Augusta, Georgia. In 1902 the system, now known as the Atlantic Coast Line (2,100 miles), was nearly doubled in length when it acquired the Plant System, a network of fourteen railroads in Georgia, Florida, and Alabama put together by Henry Bradley Plant (1819–99). About the same time, John Skelton Williams (1865–1926), a young Richmond banker, completed the organization of the Seaboard Air Line (2,600 miles), a system whose route basically paralleled that of the Atlantic Coast Line from Virginia to Florida. In 1900, with 2,300 miles of its extensive system located south of Cairo, the Illinois Central was also a major southern railroad.

Nearly as significant as the physical extension of the railway network was the greater integration and operating efficiency achieved by the adoption of a host of technological improvements in the generation after the Civil War. The existing rail network in 1865 was certainly not integrated or really very efficient. The lack of physical rail connections in several cities and the absence of bridges at wide rivers kept the system from being a complete network. The diversity of gauge, especially in the South, also made impossible the cross-country shipment of freight without break of bulk. Manual braking and coupling of the cars made railroad operation both dangerous and slow. Iron rails, wooden bridges, light locomotives, and small-capacity cars kept down the volume and speed of freight shipments. The crazy-quilt pattern of hundreds of different local times and inadequacies in train control and signaling also slowed rail traffic.

Much of this was to change in the quarter of a century after the war. As the American economy gradually shifted from a merchant to an industrial and finance capitalism, more and more American businessmen changed their thinking about railroad service. Railroads ceased to be thought of as agencies to serve the provincial or regional needs of a particular city and were viewed instead as facilities which, once co-ordinated, might serve larger national needs. Such a shift in the function and role of the railroad was obviously needed as the American economy grew with almost explosive force in the late nineteenth century. Behind Great Britain, Germany, and France in both coal and pig-iron production at the outbreak of the Civil War, the United States had easily moved into first position before 1900. As the American rail network became a more unified and efficient transportation system, it significantly contributed to this new industrial dominance.

One of the first steps toward a more integrated rail system came at the close of the Civil War when a number of railroads participated in fast-freight lines. The fast-freight line offered the shipper the advantage of greater speed between two major cities without the inconvenience of having to break bulk at normal transfer points. The latter feature was possible because the fast-freight lines, as independent stock companies, owned their own rolling stock and could operate their special trains over the trackage of several different railroads. Serious abuses often appeared in the new lines, some of which had started to operate even before the

Civil War. The new fast service tended to draw to itself the high-class, higher-rate freight, leaving to the railroad only the lower-class, less profitable business. Also, since railroad officials were frequently financially involved in the new freight companies, a dualism of control developed, which brought criticism from ordinary railroad stockholders, as well as occasional inquiries by a legislature or railroad commission.

These abuses were largely overcome with the appearance in 1866 of "co-operative" fast-freight lines. The co-operative line, unlike the original fast-freight lines, was not a separate company but simply an administrative pool, with each of the participating roads contributing cars according to its length or share of traffic. Soon every major trunk road was in one or more of the new combinations. By 1874 they were carrying the bulk of the nation's through railroad freight, and eventually there were probably forty different lines. The New York Central was in the Red, White, and Blue lines; the Pennsylvania worked with the Star Union, the National, and the Empire lines; the Great Western Dispatch was used by the Erie; while the Baltimore & Ohio reached St. Louis via the Continental Fast Freight Line. South of the Ohio the Green Line, organized in 1868, had twenty-one member railroads by 1873. All the new co-operative lines offered shippers through bills of lading. They were so efficient that by the middle seventies the bulk of the grain traffic to Boston had shifted from water to rail. Competition between the new lines grew in the eighties. In 1891 Boston was served by thirty-one fast-freight lines and Chicago by more than twenty.

In the sixties fast freights, slow freights, and even passenger trains, were still plagued by a pair of devices, the hand brake and the link-and-pin coupler, which kept all rail service slow and hazardous. The link-and-pin coupler was so arranged that the brakeman had to stand between the cars in order to steer the link into the socket and drop the pin. This job was so dangerous that brakemen were often recognized by missing fingers or a crippled hand. Often a yardmaster, when interviewing a boomer brakeman for a job, would be satisfied that the applicant was an "old-timer" if he could show one or two missing fingers. Manual braking of the cars was equally hard on life and limb. Twisting hand brakes on the top of moving cars, especially in an icy blizzard at night, was not an easy task, even for experienced trainmen.

Dozens of inventors, with hundreds of patents, sought to build a coupler which would close on impact yet open from the side of the car. The first effective automatic coupler was invented by a Confederate veteran just after the war. Major Eli H. Janney (1831–1912), while clerking in a dry-goods store near Alexandria, Virginia, used his penknife to whittle out his first model coupler. He patented his device in 1868, improved it by 1873, and managed to get the Pennsylvania Railroad interested shortly thereafter. However, few orders came to his Janney Car Coupling Company during the next few years.

In the same years, a New Yorker, George Westinghouse (1846–1914), was working on the problem of brakes. Westinghouse, getting his idea from French tunnel engineers who were cutting rock with compressed air, patented his air brake in 1869 when he was but twenty-two. The first test of his new invention, made that same year on a short passenger train near Pittsburgh, seemed a dramatic success when a farmer's wagon that was stalled on the tracks was saved from what seemed to be certain destruction. But railroad men were hard to convince. Commodore Vanderbilt, when importuned by the young inventor, roared: "Do you pretend to tell me that you could stop trains with wind? I'll give you to understand, young man, that I am too busy to have any time taken up in talking to a damned fool." Westinghouse persevered, however, improved his device in 1872 by adding a triple valve providing air pressure to each car, and managed to get his invention placed on a few eastern passenger trains. The inventor early saw the need for a standard interchangeable air-brake apparatus. Finding the railroad industry unwilling to provide him with extensive orders, Westinghouse moved on to other things. He became interested in railroad signaling, secured 125 patents in a variety of fields in the eighties, and in 1886 moved into the electrical field by organizing the Westinghouse Electric Company.

It was an Iowa farmer who was largely responsible for the final widespread acceptance of the inventions of Janney and Westinghouse. Lorenzo S. Coffin (1823–1915), former teacher and chaplain to the 32nd Iowa Infantry during the war, became appalled at the loss of life and limb suffered by trainmen and brakemen in the seventies. When he discovered that many railroad managers were refusing to use the new couplers and air brakes simply because they were expensive while labor was cheap, he became a fanatic for reform. Rebuffed by rail officials, he traveled widely, spoke frequently, and wrote endlessly—all for the cause of railroad safety. As railroad commissioner for Iowa (1883–88) he arranged for air-brake tests by the Burlington Railroad near Burlington, Iowa, in 1886–87 which eventually proved that Westinghouse's invention was effective, even for long, fast trains going downgrade. Soon Iowa passed state legislation requiring all trains in the state to use automatic couplers and air brakes. About the same time, the Master Car-Builders Association, after some 3,000 patents had been issued for coupling devices, adopted a standard model based on Janney's patents. The coupler worked like the hooked fingers of two hands, essentially being the same coupler used today. Coffin was also instrumental in pushing through to final passage the Railroad Safety Appliance Act of March, 1893. Approval of the act by President Benjamin Harrison (1833–1901) meant that all trains would be equipped with automatic couplers and air brakes. The improvement in railroad safety was dramatic. The railroad-employee accident rate, which had gone as high as 30,000 in 1881, was quickly cut 60 per cent.

Standardization touched the railroad in other ways in the eighties, when both standard time and uniformity of gauge were adopted during the decade. Standard automatic couplers and air brakes could do little to speed or facilitate through freight traffic as long as a diversity of gauge persisted. In 1861 more than 46 per cent of the nation's rail mileage was other than the 4 feet 8½ inch "standard gauge." While the use of standard gauge in the building of the Pacific railroads, and such shifts to standard as that completed by the Erie in 1880, helped bring gauge uniformity, the latter year still found a fifth of the national mileage to be other than standard.

Several expedients were used in the sixties and seventies to permit the interchange of equipment between lines of different gauge. The "compromise car" was the simplest. Having wheels whose tread was five inches wide, the cars could be used on either standard-gauge track or track as wide as 4 feet 10 inches. However, careful railroad operators frowned upon the use of such cars because they claimed that many accidents could be traced to them. A second innovation, the car with wheels that could be made to slide along the axle, was no safer and was never widely adopted. Car hoists, or "elevating machines," with the car lifted to a set of trucks of different gauge, were much safer and were used extensively by such lines as the Erie and the Illinois Central (for its service south of Cairo). A number of lines also went to "double gauge," the addition of a third rail, permitting the use of equipment of different gauge.

There was no substitute for the adoption of a single gauge by the whole nation. In the early eighties most of the gauge divergence was found in the narrow-gauge lines of the mountain West and in the Old South, where the five-foot-gauge mileage had actually increased from 7,300 miles during the Civil War to more than 12,000 miles in 1880. The Chesapeake & Ohio, the Illinois Central, and the Mobile & Ohio had all changed to the narrower standard gauge by the middle eighties. James C. Clarke, general manager of the Illinois Central lines south of Cairo, spent weeks of careful preparation for his change of gauge in the spring and early summer of 1881. On July 29, 1881, between dawn and 3:00 P.M., 3,000 workers shifted the gauge on the entire 550-mile line.

The rest of the South soon gave in. Representatives of southern lines totaling more than 13,000 miles agreed early in February 1886, to change their gauge the following May 31 and June 1. During the weeks before the day of the change, part of the southern rolling stock and motive power was changed to the narrower gauge and track gangs moved alternate inside spikes on one rail to the new position. The actual shift was but three inches, since the southern roads had voted to match the 4 foot 9 inch gauge of the Pennsylvania, the northern line with which they had the greatest interchange of cars. Using track gangs of from three to five men per mile of track, ten roads west of the mountains shifted their rail on the last day of May. The remaining roads shifted June first. On both days the work was accomplished between 3:30 A.M. and 4:00 P.M., during which period all traffic was stopped. Southern railroads had truly become part of the national network, and passenger or freight cars could move from any southern depot to any part of the nation without change of trucks or bulk. Quietly, in the next decade, all roads with a 4 foot 9 inch gauge shifted half an inch to the official standard gauge required by the American Railway Association.

An immediate dividend of increased operating efficiency resulted from gauge standardization as a system of more extensive car interchange developed among the railroad companies. For years railroads had permitted carload freight to be shipped through to the consignee, but the fast-freight lines were still handling much through freight as late as 1890. The interchange of cars was accelerated as a rental system of "car mileage" (0.75 cents per mile for several years) was introduced. This type of car rental was not very satisfactory because many railways used foreign cars for storage purposes. In 1902 a per diem system of 20 cents per day per car was adopted. Per diem charges were raised several times, being increased to $1.00 per day in 1920, a rate which was used for many years. Inflation in mid-century brought further increases, and late in 1959 it was raised to $2.88 per day. Early in the century an extensive code of car-service rules was worked out to secure an early return of foreign cars to the home road and to assure an equitable division of the costs of car repair.

In the post–Civil War generation vast improvements were made in the rail. The first steel rail was used in the United States in 1863, when the Pennsylvania imported 150 tons from England. In May, 1865, the North Chicago Rolling Mills produced the first domestic-made steel rails, but acceptance was slow and the total national production of 2,277 tons in 1867 was enough for only a few miles of line. However, the advantages of the long-wearing new rail were obvious, especially since the price was low. By 1880 the annual production had passed the one-million-ton mark and perhaps a quarter of the nation's mileage was steel. In 1890 the figure had increased to 80 per cent, and virtually all mileage was laid with steel by 1910. The rail also became heavier during the years. Much of the steel rail rolled in 1880 weighed but 35 pounds to the yard. By 1920 the average for all railroads in the United States had risen to 82 pounds per yard. And rails also increased in length. The standard rail lengths of 30 feet and 33 feet of the late nineteenth century were pushed up to 39 feet in 1925.

Standard time was adopted during the eighties. As the railroads spanned the nation and became a national network after the Civil War, they ran their trains with a crazy-quilt pattern of dozens of times based upon several mean local sun times. When it was noon in Chicago, it was 11:27 A.M. in Omaha, 11:50 in St. Louis, 12:09 P.M. in Louisville, 12:17 in Toledo, and 12:31 in Pittsburgh. In fact, in Pittsburgh there were six different times for the departure and arrival of trains. The train station in Buffalo had three clocks, each with a different railroad time, while the *Chicago Tribune* claimed there were 27 local times in Illinois and 38 in Wisconsin. The Union Pacific operated trains on its extensive system with at least six different local sun times. In Kansas City, where the leading jewelers furnished "standard time," sometimes with a variation of as much as twenty minutes, the confusion was untangled only when a single "time ball" was dropped each noon to indicate a uniform time.

As early as 1870 the *Railroad Gazette* had urged the adoption of a single, standard time zone for the entire nation, a proposal which received little support. Professor C. F. Dowd (1825–1904), who ran a young ladies' seminary at Saratoga Springs, New York, was one of the first to suggest that the country be divided into four or more broad time zones or belts. It was William F. Allen (1846–1915), managing editor of the *Official Guide of the Railways* and secretary of the General Time Convention, who finally convinced the lines that they should adopt a standard time. His plans, as conceived by 1881, were accepted by the railroads in October, 1883, to go into effect at noon on Sunday, November 18, 1883. Zones based on the 75th, 90th, 105th, and 120th meridians divided the nation into four time belts. As the shift in time took place, people living in the eastern half of each zone experienced the "day of two noons," while people farther west were thrown abruptly into the future. Public adoption of the new railroad time was general, although one editor did complain that he would rather run his clock on "God's time—not Vanderbilt's." Just before the change, the Attorney General righteously declared that no government department need use the new time until so authorized by Congress. The same gentleman, according to a doubtlessly apocryphal story, was astonished when he missed a late afternoon train on the eighteenth by eight minutes and twenty seconds. Congress officially adopted standard time thirty-five years later, in 1918.

As standardized equipment and uniformity in operation resulted in faster more efficient railroad service, improvements in the signaling and control of trains became necessary. A disastrous rear-end collision between two troop trains in New Jersey moved Ashbel Welch (1809–82), vice-president and engineer of the Camden and Amboy, to develop in 1865 the first manual block-signal system in America. This idea of keeping trains apart by a certain interval or block of distance was improved upon in 1871 when William Robinson (1840–1921) devised a closed electric track circuit to set the signals. In the same decade the control of train movements at junctions and terminals was vastly improved when Welch and others introduced manual mechanical interlocking, an arrangement which made it impossible for signalmen to line up switches and signals in conflict with each other. Pneumatic and electrical interlocking soon followed. The basic principles of block signals, interlocking, and electrical track circuits were to be refined later, in the twenties, into systems of automatic train control (required on certain lines by the ICC in 1922) and centralized traffic control, where a single man at a control desk can manage switches, signals, and train movements over an entire division.

Naturally, if there were to be longer and heavier trains in the generations after the Civil War, significant improvements were also required in railroad motive power. Several improvements in locomotives had developed during the Civil War. In 1862 Matthias Baldwin imported five hundred steel tires and began the practice of shrinking them on the iron locomotive wheels. About the same time, successful experiments with steel fireboxes and boilers permitted higher steam pressure and more power. By 1865 the new steam injector was replacing the older boiler pump. In the interest of both beauty and improved performance, most new locomotives had their cylinders in a horizontal position instead of on the diagonal. Changes in fuel were also being made. Exhaustive tests by the Pennsylvania Railroad just before the war indicated that in heating

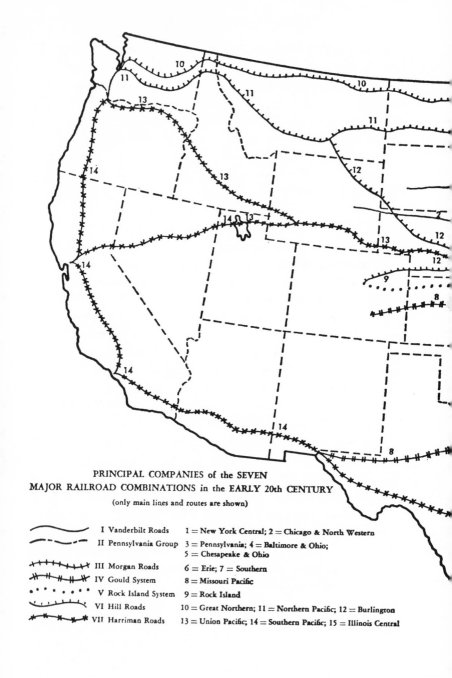

PRINCIPAL COMPANIES of the SEVEN
MAJOR RAILROAD COMBINATIONS in the EARLY 20th CENTURY

(only main lines and routes are shown)

	I Vanderbilt Roads	1 = New York Central; 2 = Chicago & North Western
	II Pennsylvania Group	3 = Pennsylvania; 4 = Baltimore & Ohio;
		5 = Chesapeake & Ohio
	III Morgan Roads	6 = Erie; 7 = Southern
	IV Gould System	8 = Missouri Pacific
	V Rock Island System	9 = Rock Island
	VI Hill Roads	10 = Great Northern; 11 = Northern Pacific; 12 = Burlington
	VII Harriman Roads	13 = Union Pacific; 14 = Southern Pacific; 15 = Illinois Central

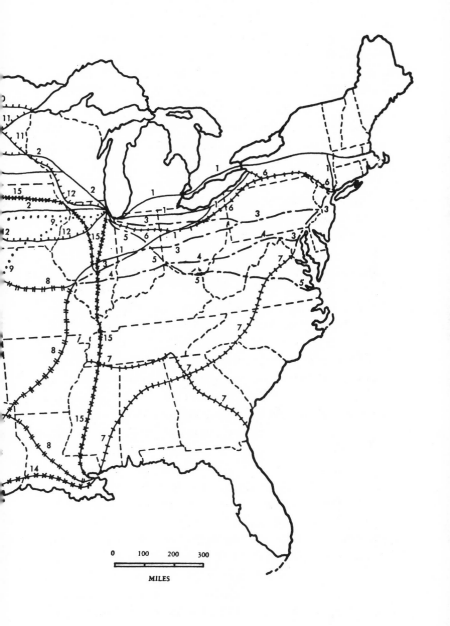

value a ton of soft coal was equal to one and one-third cords of hard wood. In 1862 the Pennsylvania withdrew the last of its wood-burning freight engines, and many other roads soon did likewise. By 1870 both the Pennsylvania and the New York Central were experimenting with track tanks, an arrangement in which a scoop in the tender was lowered to take up water from a pan between the rails while the train was moving at a moderate speed.

The major motive-power development after the war was in the size of the locomotive. In New England, where the heavy passenger traffic and moderate freight business permitted it, the railroads clung to the proved American-type engine. There the Taunton, Massachusetts, locomotive works of William Mason (1808–83), a cotton-machinery expert who turned to locomotives, produced hundreds of clean-cut engines noted for both their beauty and excellence of workmanship. The American-type locomotive is known as a 4-4-0 (four wheels in front, four drivers, no wheels under the cab) according to the Frederic M. Whyte (1865–1941) system of engine classification. Suburban traffic, and later the elevated lines of New York City and Chicago, also used small locomotives, especially the Forney engine, a tank locomotive invented in 1866 by Matthias N. Forney (1835–1908), an inventor who had once been an apprentice to Ross Winans. Most western roads in the late nineteenth century were trying heavier locomotives with more and smaller driving wheels. By the Civil War years the master mechanics and engineers of the Baltimore & Ohio were shifting from the Ross Winans camel-backs to the newer ten-wheelers (4-6-0). The "Thatcher Perkins," built in 1863 and weighing, with tender, nearly 90,000 pounds, was one of the best known ten-wheelers on the B.&O. By 1865 many lines were also building the Mogul locomotive (2-6-0), which provided still greater traction by having 85 per cent of the engine weight on the drivers.

The problem presented by locomotives with long wheelbases in negotiating curves was solved by using flexible-beam driver trucks (invented in 1842 by Baldwin), by providing the journal boxes with lateral play, by removing the flanges from the middle drivers, and by slightly increasing the track gauge on curves. When the Lehigh Valley, a consolidation of a number of short lines, ordered (in 1867) a new type of locomotive with a fourth pair of drivers, still more tractive force was achieved. The new engine, called Consolidation (2-8-0) in honor of the Lehigh Valley, became the standard heavy freight engine for the next quarter of a century. By 1890 the American and ten-wheeler engines used in passenger service weighed, with tenders, as much as fifty tons. The Consolidations and Decapods (2-10-0) which provided freight service weighed up to seventy-five or eighty tons.

Little additional power could be expected without a larger firebox. This was achieved in the nineties as the firebox was taken from between the rear drivers, made much broader, and placed over a supporting trailing truck. For passenger service the small Columbia (2-4-2) appeared in 1892, the four-hundredth anniversary of the discovery of America. Three years later the Atlantic type (4-4-2), an American with a trailing truck, was built for the Atlantic Coast Line. By the turn of the century the Missouri Pacific had ordered the first Pacific (4-6-2). Early in the twentieth century Americans saw their first articulated, or Mallet, locomotive, an engine with a single boiler and firebox driving two sets of drivers designed by the Frenchman Anatole Mallet (1837–1919). The Baltimore & Ohio ordered a Mallet (0-6-6-0) in 1904, and in 1913 the Erie reached a peak in wheel complexity with a triplex Mallet (2-8-8-8-2). But such innovations as superheated steam and the automatic stoker were more important to most railroads in the early twentieth century.

Freight equipment increased in size and quality along with motive power. As train loads of from one hundred to two hundred tons increased to two thousand tons or more in the half-century after the Civil War, average freight-car capacity increased from about ten tons to about forty tons per car. Some of the first improvements in freight equipment accompanied the introduction of the fast-freight lines. A greater variety of special cars also appeared shortly after the war. Some cattle had been moved by rail before the war, but the major appearance of special stock cars came with the growth of the cattle business in the seventies and eighties. The care of the cattle en route to market was much improved in the early eighties with the addition of feed bins and watering troughs to the cars and the patenting of such new models as the Mather Palace Stock Car by Alonzo Mather (1848–1941), a young Chicago businessman who disliked seeing inhumane treatment of livestock.

Crude attempts at refrigeration, using ice and sawdust in boxcars, had been tried before the war, but the first patent for a refrigerator car was issued in 1867, the same year that the Blue Line added such service to its fast freights. The Illinois Central had fast, refrigerated strawberry trains in the seventies and comparable service for bananas a decade or so later. Meat-packers also used the new cars extensively, generally painting their private cars to advertise their product. The first oil shipped by railroad out of the new oil region in Pennsylvania left Titusville in 1865 in two large wooden tubs fastened to a flat car. In 1868 the present type of horizontal tank car fitted with a dome for expansion was introduced. Coal-carrying cars had been used long before the war, but by 1880 hopper cars with all-metal bodies were being built, and the capacity of some coal cars had been increased to fifty tons by the late nineties. The cabin car, or caboose, for the conductor and brakemen also became standard before the war. It gained its distinctive feature, the cupola, during the Civil War when a resourceful conductor on the Chicago and North Western rigged up a seat with a skylight above the roof of his car. While it could hardly be classed as a normal freight movement, the first use of the railroad by a circus came in 1872 when Phineas T. Barnum (1810–91) exchanged his six hundred draft horses for sixty-five flamboyant railway cars to carry his "Museum, Menagerie and Hippodrome" from town to town.

In the generation after the Civil War there was a much smaller increase in the size and capacity of passenger equipment than had occurred in freight facilities and motive power. There were definite innovations and improvements in the period, however, both in the variety of cars and in the comfort and safety afforded the passengers. Several railroads had provided sleeping cars in the years before the Civil War, but the service was so crude and uncomfortable in the hard, narrow berths that passengers rarely bothered to undress or remove their boots. In 1864 and 1865 George Pullman decided to improve upon the earlier cars he had built (before the war) for the Chicago and Alton. His "Pioneer," which cost $20,000, when no previous car of any type had cost more than $5,000, was enough larger than existing equipment that it would cause many lines a clearance problem. Finished with hand-carved woodwork, plush carpets, and fine mirrors, the new car was first used when Mrs. Lincoln asked if it could be added to the President's funeral train for the trip from Chicago to Springfield. The Chicago and Alton promptly made the necessary changes in structures, and the "Pioneer" was used for the first time.

Soon the "Pioneer" and comparable cars were being used, at premium rates, which the public gladly paid, on the Alton, the Michigan Central, the Burlington, and the Great Western. When Pullman introduced sheets and bed linen as standard equipment, he had some trouble. The request "Please Take Off Your Boots Before Retiring" was placed on placards in every car and printed on every ticket. In 1863, two day-coaches had been converted into primitive dining cars by the Philadelphia, Wilmington, and Baltimore for service between Baltimore and Philadelphia. Pullman gave dining service the de luxe treatment in 1868 when he built the "Delmonico," named after the restaurateur.

Always an effective publicist, Pullman in 1867 had shown his latest cars to the traveling public with a special excursion from New York to Chicago. In the late spring of 1870 he outdid himself in providing a 6,000-mile transcontinental round trip "From Faneuil Hall to the Golden Gate" for 129 members of the Boston Board of Trade. The special train consisted of eight "of the most elegant cars ever drawn over an American railway," all Pullman built of course. In the years that followed, Pullman produced hundreds of sleepers, parlor cars, and dining cars in the newly built "company town" of Pullman on the outskirts of Chicago. Pullman operated his cars so efficiently that rival companies, such as those of Edward Knight, T. T. Woodruff, and Webster Wagner, were soon forced to sell out to him.

As the Pullman cars grew in elegance and ingenuity, other passenger equipment was improved. Kerosene lamps and, later, the Pintch gaslight from Germany replaced candles in most passenger equipment in the sixties and seventies. In 1882 the Pennsylvania tried lighting a standard passenger car with electricity, and entire trains on a number of roads were thus illuminated by 1887. Hot-water heaters replaced stoves in some cars in 1868, and steam heating from the locomotive was introduced by 1881. Six-wheeled trucks for passenger cars were available by 1876, and in 1879 the steel-tired wheel was ready to replace the older cast-iron wheel. Steel underframes, center sills, and platforms were available by the end of the century, and in 1906 all-steel passenger coaches were in regular service on the Pennsylvania and the Southern Pacific.

Before the use of diners, railroads had tried to prevent their passengers from moving from car to car across the wind-swept, swaying open platforms of moving trains. Since movement between cars now became necessary, great attention was given to increasing passenger safety. The answer to the problem came in 1887 with the introduction of the vestibule (a flexible, covered passageway between cars) devised by H. H. Sessions (1847–1915), master car-builder and superintendent of the Pullman Company. Solid vestibule trains were soon in service between New York and Chicago on the Pennsylvania Railroad. Full car-width vestibules replaced the original narrow door-width models after 1893. Both types had safety advantages through the reduction of shock and sway and the reduced chance of the cars' telescoping in the event of a collision. During the first half-century of its existence, the Pullman Company also built hundreds of private cars for the company presidents and tycoons of the nation, cars that were the last word in elegance, luxury, and refinement.

With all the improvements and innovations, railroad travel was still fairly dangerous in the last generation of the nineteenth century. Massachusetts, whose records were perhaps the most complete during the period, reported that her railroads killed (passengers, employees, and trespassers), on the average, 87 people per year in the sixties, 143 per year in the seventies, and 208 per year in the eighties. The accident rate was even worse for the rest of the nation. But Charles Francis Adams, Jr., did not find the accident rate really excessive, writing in 1879 of the railroad: "A practically irresistible force crashing through the busy hive of modern civilization at a wild rate of speed, going hither and thither, across highways and by-ways . . . cannot be expected to work incessantly and yet never come in contact with the human frame." Come in contact they did, as when a rear-end collision at Revere, Massachusetts, killed 29 and injured 57 in 1871; a holocaust at Ashtabula, Ohio, cremated dozens as the crack Pacific Express on Vanderbilt's Lake Shore & Michigan Southern crashed through a bridge in 1876; and another wreck in 1887 at Chatsworth, Illinois, killed more than 80.

These and other accidents resulted from many things. The Revere tragedy, the ensuing lawsuits of which nearly bankrupted the Eastern Railroad, was clearly the result of a rail management so inept and old-fashioned that in 1871 it was still refusing to use the telegraph for dispatching trains. The Ashtabula and Chatsworth disasters, both of which deeply shocked the country, were caused by a combination of kerosene lamps, hot car stoves, and defective bridges. Fire figured in almost every wreck with a long casualty list. Railroad bridges in the sixties and seventies were still frequently made of wood and often too lightly built for increasingly heavier traffic.

At the time of the Civil War the standard railroad bridge was the Howe truss, a rectangular trussed frame of wooden diagonals and vertical iron tie rods. Invented by William Howe (1803–52), farmer turned inventor and uncle of Elias Howe, the new truss was an improvement of an earlier bridge form devised by Colonel Stephen H. Long (1784–1864) of the Army Engineers. The wreck at Ashtabula was due to the failure of an all-iron Howe truss bridge built in 1863 at the insistence of the road's president, Amasa Stone (1818–83), and against the advice of a subordinate engineer. Stone, brother-in-law and long-time bridge-building partner of William Howe, was proved wrong thirteen years later. Such accidents as that on Stone's line hastened a new emphasis on the building of iron and steel railroad bridges with proved margins of safety. The campaign for safer train operation paid off. Railroad passenger safety increased threefold between the early nineties and the years just before World War I, by which time there was far less than one fatality for each one hundred million passenger-miles of service.

In the last third of the nineteenth century, train robberies, while less frequent and dangerous than wrecks, generally caused equal excitement. The gun in the ribs of the baggage man, the pilfered strongbox in the express car, the tossing of passengers' watches, cash, and valuables into the renegade's grain sack—all became a possible hazard of train travel. The activities of the rail desperado also loomed large in the dime-novel literature and folklore of the period. The Reno gang of Indiana was the first well-organized band of train robbers. After successfully holding up several trains on the Mississippi and Ohio Railway and other roads in the late sixties, the activities of the Reno boys were permanently stopped by some indignant vigilantes well supplied with rope. The crimes of two Missouri-born outlaws, Jesse James (1847–82) and Thomas Younger (1844–1916), were a little harder to stop. Both James and Younger had served some sort of an apprenticeship for the life of a desperado as members of the Quantrill band during the Civil War. As the James-Younger gang robbed train after train in Iowa, Missouri, and Kansas in the middle seventies, they were aided by a public opinion that was tolerant, if not sympathetic. Farmers who felt cheated by high freight rates, investors who had lost in wildcat rail investments, railroad workers who remembered the strike violence of 1877—all were little worried if other men successfully robbed the cars. Two other bandits, Sam Bass (1851–78) and Robert Dalton (1867–92), both of whom started out as deputy marshals, were cut down in their middle twenties as they preyed upon southwestern trains and banks. There was a decline in rail banditry by the nineties, and after the turn of the century, the steel mail-and-express car, a shift in public opinion, and better police methods combined practically to eliminate the robbing of trains.

The dominant characteristic of American railroads in the decades after the Civil War was an ever increasing efficiency of service and operation. The changes in rail, gauge, rolling stock, and motive power all contributed to operational economy. Larger engines with higher tractive effort made for economies in fuel and the use of labor. Better equipment, the increased use of steel rails, and the achievement of a truly integrated network resulted in average higher train speeds, which permitted both cars and engines to cover more mileage per year. The improvements in signaling and train control often made the laying of additional track unnecessary.

Larger capacity cars, often without any proportionate increase in dead weight, meant heavier and longer trains. The Illinois Central, for example, increased its average net train load from just under one hundred tons in 1870 to well over five hundred tons in 1915.

These great increases in operational efficiency were responsible for, and accompanied by, significant reductions in freight rates. Average freight rates declined from something over 2 cents a ton-mile at the end of the Civil War to about 0.75 cents a ton-mile by 1900. In the years of falling prices after the war, railroad rates certainly fell more rapidly than the general price structure, and as prices tipped upward early in the twentieth century, freight rates generally lagged behind. The relatively cheap rates, plus increasing rail efficiency, gave the railroads a growing monopoly of domestic transportation in the postwar generation.

Further increased by the rapid industrialization of a nation whose population was doubling each generation, American railroad freight statistics seemed to explode in the half-century after 1865. Total freight traffic grew from perhaps 10,000,000,000 ton-miles in 1865 to 79,000,000,000 ton-miles in 1890 and 366,000,000,000 ton-miles in 1916. Some indication of the increased efficiency of operation can be seen in the fact that slightly more than a fourfold increase in freight cars between 1880 and 1920 permitted an increase in freight ton-mileage during the four decades that was nearly elevenfold. In the half-century between the wars the average American relied more and more on the railroads. The per capita annual freight service increased from 285 ton-miles in 1867 to 1,211 in 1890 and 3,588 ton-miles per person in 1916.

Some of the great increase of railroad traffic in the late nineteenth century was at the expense of the nation's rivers, waterways, and canals. The Erie Canal had outlasted all the other man-made waterways but was losing out to rail competition by the end of the Civil War. The fierce railroad rate wars of the seventies proved to be the final blow. By 1876 more than four-fifths of all grain received at eastern ports came by rail. In 1878 the freight receipts at New York City arriving via the New York Central, the Erie, and the Pennsylvania stood at twenty-five million tons, five times the volume of the Erie Canal. Canal barges were generally relegated to such low-grade freight as sand, coal, and cement.

The rivers soon followed the canals in declining traffic. The decline of the western river traffic from the lush days of the fifties is most graphically shown by Mark Twain (1835–1910) in his *Life on the Mississippi*:

> Boat used to land—captain on the hurricane-roof—mighty stiff and straight—iron ramrod for a spine—kid gloves, plug hat, hair parted behind—man on shore takes off hat and says:
>
> "Got twenty-eight tons of wheat, cap'n—be great favor if you can take them."
>
> Captain says:
>
> "I'll take two of them"—and don't even condescend to look at him.
>
> But nowadays the captain takes off his old slouch, and smiles all the way around to the back of his ears, and gets off a bow which he hasn't got any ramrod to interfere with, and says:
>
> "Glad to see you, Smith, glad to see you—you're looking well—haven't seen you looking so well for years—what you got for us?"
>
> "Nuth'n'," says Smith; and keeps his hat on, and just turns his back and goes to talking with somebody else.

On the Mississippi in the seventies the passenger business was probably hurt more than the freight traffic. Still, when the "Robert E. Lee" won its famous race with the "Natchez" on the New Orleans–St. Louis run in July, 1870, there was not enough freight at the Missouri port to permit the victor immediately to return downstream. The railroads continued to get the bulk of all commercial traffic well into the twentieth century. In 1916 the Interstate Commerce Commission reported that the nation's steam and electric railways were carrying more than 77 per cent of the inter-

city freight traffic, and 98 per cent of the intercity passenger business. Most of the non-railroad traffic was on the Great Lakes
rather than on rivers and canals.

While railroads possessed a full monopoly of the commercial
passenger business in the early twentieth century, this portion of
the rail traffic had not increased as rapidly as the freight business
in the decades since the Civil War. Passenger miles per year had
increased from nearly five billion in 1870 to twelve billion in 1890
and thirty-five billion in 1916. This sevenfold increase in a short
half-century was considerably smaller than the freight increase because passenger traffic lacked the elasticity of freight business, especially during the period of rapid American industrial growth. As
a result, passenger revenue in the half-century dropped from a
large fourth of the total rail revenue to roughly a fifth. This decline
in the relative importance of passenger revenue was slowed by the
fact that passenger fares in the period dropped more slowly than
freight rates, being reduced on the average from something over
three cents a mile in 1865 to two cents a mile in 1916.

Passenger traffic was greatly facilitated in the postwar years as
iron and steel bridges were built across major rivers. In 1865 the
North Western finished a bridge across the Mississippi (the second
south of St. Paul) at Clinton, Iowa, and in 1868 the Burlington
completed another at Burlington, Iowa. Six years later Captain
James B. Eads (1820–87) completed his bridge at St. Louis. The
Missouri River was crossed at Kansas City in 1869 and at Omaha
in 1871. Both Cincinnati and Louisville completed spans across
the Ohio in 1870, and the Illinois Central finally bridged the same
stream at Cairo in 1889. The substitution of bridges for car ferries,
such as the steamers used at Cairo in the seventies, permitted further increases in passenger-train speed. American-type locomotives had always been noted for their speed, and it was an engine
of that type, New York Central "No. 999," which first attained a
speed of over 100 miles an hour, running a mile in 32 seconds

(112.5 miles per hour), near Batavia, New York, on May 10, 1893. A dozen years later on the Pennsylvania the "Pennsylvania Special," near Ada, Ohio, ran three miles at the rate of 127 miles per hour. In the same year, 1905, a rich eccentric from California, "Death Valley Scotty," paid the Santa Fe $5,500 for a record run of 2,265 miles to Chicago. The special train reached Dearborn Station, Chicago, in 44 hours, 54 minutes, beating the best previous time by more than 13 hours.

The desire for more rapid movement of the mails also helped speed up passenger traffic. In the summer of 1862 the first experimental mail car for the sorting of mail en route was placed in service between Hannibal and St. Joseph, Missouri, which had been the eastern terminal of the Pony Express. By the end of the decade mail clerks were sorting mail on trains in most of the country. In the middle seventies George S. Bangs (1825–77), in charge of the Railway Mail Service, suggested to the New York Central that a special, fast, all-mail train be run between New York City and Chicago. The old Commodore rejected the idea, but William H., with his father's grudging consent, built a special train of cream-and-gold mail cars. The first "Fast Mail" had a twenty-four-hour schedule, fast time for the seventies.

As the American rail network expanded and improved its services in the decades after the Civil War, the total rail investment grew rapidly. The total capitalization of the railroads expanded from $2,500,000,000 in 1870 to $10,000,000,000 in 1890 and $21,000,000,000 in 1916. In the same years, the gross railroad revenue rose from $400,000,000 in 1870 to $1,000,000,000 in 1890 and $3,500,000,000 in 1916. The depressions of the seventies and nineties had kept rail earnings uneven in the generation following the Civil War, but the dividend picture brightened by the end of the nineteenth century. Between 1895 and 1911 the proportion of railroad capital stock which paid dividends increased from 30 per cent to 68 per cent. The rate of return in the latter year was 8 per cent on the dividend-yielding stock but only 5.4 per cent on the total capital stock.

One aspect of the relative rail prosperity of the turn of the century was the building of a number of elaborate and luxurious stations and union depots. The St. Louis Union Station, built to serve 18 different railroads with its 42 stub tracks, was completed in 1894, while Boston had its new South Station by 1898. The beautiful Union Station in Washington, D.C., was built in 1907. But the greatest terminals were being built in New York City. The Pennsylvania tunneled under both the Hudson and East rivers and opened its Pennsylvania Station in 1910 at a cost of more than $100,000,000. Three years later the New York Central completed its Grand Central Terminal. Electric locomotives served both the new terminals.

Some portions of the rail financial picture were not so favorable in the early twentieth century. A fundamental weakness was the growing amount of funded debt carried by the railroads. Until the late eighties capital stock had accounted for a majority of the total rail securities. In the next three decades the rail debt increased so rapidly that by 1916 the public-held, unmatured funded debt stood at $9,900,000,000, while the public-held capital stock came to only $6,400,000,000. Another standard of railroad fiscal health, the operating ratio (the ratio of operating expenditure to operating revenue), also showed disturbing symptoms. Nineteenth-century experience had shown that an operating ratio of 65 per cent to 68 per cent indicated a fairly prosperous rail industry. Between 1890 and 1910 the operating ratio had averaged 66.6 per cent, but in the teens, even before federal operation, it had climbed to an average of almost 70 per cent. Higher labor costs contributed to the higher operating ratio. The compensation of labor per rail revenue dollar rose from an average of 40 cents in the years 1895–99 to 46 cents in the years 1911–15.

Rail employment expanded rapidly in the late nineteenth century and hit an all-time peak of just over two million workers in 1920 as railroads were handed back to private management after two years of federal control. As rail traffic volume increased, railroad employment grew faster than new line was built. The portion of the gainfully employed engaged in railroad work increased until one worker in twenty-five was thus employed in the years just prior to World War I. The decline in the number of rail workers after 1920 was the result of the twin forces of an increased efficiency in railroad operation and an ever growing competition from new transport facilities. Increased efficiency in freight service was so marked that in 1916 the average rail worker was producing about two and one-half times as much freight ton-mileage as the worker in 1880.

In the late nineteenth and early twentieth centuries the railroad worker's pay increased almost as rapidly as his productivity. His average annual wages rose in the seventies, eighties, and nineties, while price levels were stable or declining. In the first years of the new century his pay increases in general were as large as the increases in the cost of living. Throughout the period the railroad employee had wages well above the earnings of the average American, whose mean annual pay was $438 in 1890, $438 in 1900, $574 in 1910, and $708 in 1916. He was also better off than the rail workers of Europe. In 1907, when the average yearly rail wages amounted to $661 in the United States, they were $371 in Germany, $292 in Switzerland, and $260 in the United Kingdom. Among American railroad men the operating personnel were the best paid. Engineers received average daily wages of $3.68 in 1892 and $4.46 in 1908 (about twice the average daily wage for the industry). For the same years the pay of conductors was $3.07 and $3.83; firemen, $2.07 and $2.76; and brakemen, $1.89 and $2.64. Undoubtedly the early unionization of railroad workers contributed to their fairly high wage scales. Following the lead of the operating brotherhoods, who founded their four unions in the two decades after the Civil War, twenty different railroad labor organizations were in existence by 1901.

The half-century following the Civil War had been a period of advancement, not only for railroad labor, but for all facets and aspects of the industry. The rapid expansion of the national rail network in the last decades of the nineteenth century was fully matched by the technological improvements adopted by the rail carriers. The increased use of fast-freight lines and the successful experiments with air brakes and improved couplers clearly showed the American businessman that an integrated national rail network was soon to be available. The adoption of steel rail, the building of key bridges across major streams, and significant improvements in signaling resulted in faster freight service. Standard time and standard gauge, both achieved during the eighties, were major strides toward uniformity and greater operating efficiency. Throughout the period average train loads went up, and freight rates declined, as locomotives of greater power and rolling stock of greater capacity were added to equipment rosters. Most observers of the national transportation scene in the early twentieth century were agreed that American railroads were providing a shipping and traveling public with an efficient and extensive service. Between 1880 and 1916, American railroads were enjoying a Golden Age, as shown by table 6.1.

In the five decades before World War I the railroads of the country had expanded and developed more rapidly than most segments of the industrial economy. While the population had not quite tripled (36,000,000 to 102,000,000) between 1865 and 1916, the nation's rail network had grown more than sevenfold in the same years. In the half-century the annual value of manufactured products in America increased nearly seventeen times, from perhaps $3,000,000,000 to about $50,000,000,000. Annual rail freight ton-mileage increased by thirty-five times in the same years.

TABLE 6.1

AMERICAN RAILROADS IN THEIR GOLDEN AGE

Year	Mileage	Total Invest-ment (billions)	Operat-ing Revenues (millions)	Railroad Employ-ees	Average Annual Wages (Curr. $)	(1880 $)	Annual Freight Ton-mileage per Employee
1880	93,000	$ 5	$ 614	419,000	$465	$465	88,000
1890	164,000	9	1,006	749,000	572	602	101,000
1900	193,000	11	1,372	1,018,000	567	591	138,000
1916	254,000	21	3,353	1,701,000	886	620	215,000

Source: John F. Stover, "Railroads," p. 908, in *The Reader's Companion To American History* (1991), with permission from Houghton Mifflin Company.

While the total national income was experiencing a sevenfold increase and the national wealth was growing ten times, annual gross railroad revenue had a twelvefold increase, from $300,000,000 to $3,600,000,000. American railroads were clearly taking care of the nation's transportation needs, even though those needs were growing with explosive speed. In the years between the two wars the American economy had developed from an agrarian-oriented second-rate industrial society into the industrial giant of the world. The nation's railroads had kept pace with the change.

As the American network of steel rails was completed in the first years of the new century, the dominance of the railroad in the carriage of both freight and passengers seemed complete. These years of completion and expansion saw the passing of the last of the great railroad-builders. Collis P. Huntington, thrifty, hard-working Connecticut Yankee to the end, died in 1900, leaving behind him the plain, cramped little office (on the seventh floor of the Mills Building in New York City) in which he had worked. Harriman died in 1909, and the elder Morgan passed on four years later. The "Empire Builder," James Jerome Hill, died in 1916 (the year of peak rail mileage and also the year of the first federal highway construction act in decades). The last of the nineteenth-century railroad-builders were gone. Their successors were to face new problems.

7

Railroads at War

Twice in the twentieth century American railroads faced the rigors of total war. Twice the railroads helped the nation to victory. Half a century earlier, before they had completed the bulk of their rail network, the railroads had played a vital role in a conflict fought at home, the Civil War. In 1917, only a few months after the lines had been built to a record peak of mileage, the railroads were called upon to meet the unprecedented transportation demands of a domestic economy preparing for total conflict. They were also expected to help supply the Allies in Europe, who seemed to have an insatiable appetite for more and more war matériel. Some of the transport lessons learned in a simpler war fifty years before had been forgotten, but the transportation crisis at Christmas time, 1917, was mainly the result of other factors: the very dependence of the nation on rail transport, the unique dimensions of war needs, and a series of unfortunate developments in the preceding years. Faced with crisis, the United States Railroad Administration took over the lines, operated them as a single system, and met the basic transportation requirements. Twenty years later, when war again came to America, the railroads remembered the unpalatable years of federal operation and resolved to do their job by themselves. Even though weakened by the depression of the thirties, the rail lines, recalling lessons learned earlier, went on to meet the tremendous transportation requirements of World War II without the benefit of any substantial direction by the government.

In the five decades after the Civil War, American railroads had made great advances in mileage, uniformity, and the general efficiency of their operation. Nevertheless, they were experiencing one of the most difficult financial periods in their history in the years just before America became involved in World War I. A major cause of the financial troubles in the teens was the steady rise in the costs of railroad operation, without any corresponding increase in railroad rates and earnings. Between 1900 and 1915 the general price level in the country went up about 30 per cent. Rail costs for fuel, taxes, and labor increased at an even higher rate. Average railroad wages increased 50 per cent in the period. Railroad taxes tripled in amount, while gross earnings only doubled in the fifteen years. Throughout the decade and a half, freight rates remained virtually unchanged at 0.73 cents a ton-mile.

The Hepburn Act of 1906, in giving the ICC the authority to establish reasonable maximum rates, had transferred the burden of proof to the railroads in future rate cases. Railroad management, which had been building up public ill will for decades, received little sympathy from either the public or the ICC as it struggled against the rising costs of operation in the early twentieth century. Requests for higher rates in 1910 were unanimously rejected by the Commission. A second request in 1913 eventually resulted in a modest 5 per cent advance in freight rates, but the relief was inadequate and was soon nullified by further hikes in operating costs. As rail traffic declined in 1914, the operating ratio rose to more than 72 per cent and the financial condition of several weaker systems became desperate. Soon the Rock Island, the Frisco, the Wabash, and the New Haven systems were faced with bankruptcy. By early fall of 1915, a sixth of the rail network of the nation, more than 40,000 miles of road with a capitalization of $2,250,000,000, was under court control or awaiting receivership. In the teens the declining credit position of the industry made it difficult for the railroads to continue planned improvements in their facilities.

Railroad traffic definitely increased in 1915 and 1916, but the returning prosperity was accompanied by new demands from labor. In 1916 the four operating brotherhoods (engineers, firemen, conductors, and trainmen), already the best paid of railroad labor, demanded an eight-hour day instead of the ten-hour day then in effect. Since the hours of train service could not easily be shortened, the demand was really a request for a major pay boost. When negotiations faltered, President Woodrow Wilson (1856–1924) suggested a compromise which included the eight-hour day without punitive overtime pay. The compromise was unacceptable to the railroad presidents, some of whom may have welcomed the prospect of a general strike as a way of breaking the unions. When the rail executives rejected the President's final plea at a White House conference on August 21, 1916, Wilson left the room, exclaiming bitterly: "I pray God to forgive you, I never can."

The brotherhoods called a nationwide strike for September 4, 1916. Fearing a national crisis, Wilson urged Congress on August 29 to pass legislation providing for the eight-hour day. Congress quickly responded with the Adamson Act, sponsored by William C. Adamson (1854–1929), chairman of the House Interstate Commerce Committee and long-time friend of railroad labor. The law gave the brotherhoods the eight-hour day, effective January 1, 1917. The railroad companies refused to obey the legislation, which they contended was unconstitutional, and sought satisfaction in the federal courts. Unwilling to wait for judicial decision, union leaders called a strike for March 19, 1917. On the morning of the nineteenth the railroad managers decided to yield, and that afternoon the Supreme Court, in a five to four decision, upheld the eight-hour law. The labor leaders had picked the best possible time to do battle for their shorter work day: rail revenues were high with a glut of defense traffic, skilled labor was hard to keep because of the high wages paid in defense plants, and the nation was on the verge of war. Two days after the decision of the Supreme Court, President Wilson issued a call for the special session of Congress that was to hear his war message.

Many problems faced the nation's rail lines as America entered World War I in April 1917. There was no shortage of rail traffic. As a Europe at war sought more and more American foodstuffs and military supplies, railroad ton-mileage increased so rapidly that freight traffic in 1917 was 43 per cent greater than that of 1914–15. As operating costs, especially those of labor, continued to rise, the railroads appealed to the ICC for a freight-rate advance of 15 per cent. The Commission replied with token increases. By the end of the year operating costs were such that the full requested rate increase, even if granted, would have been inadequate. Efficient railroad operation was also difficult during the war because the bulk of all freight shipments was headed in one direction: toward a few eastern seaports. The successful German submarine campaign of 1917 reduced the shipping tonnage available in these ports, and ports and rail terminals alike became glutted with goods waiting for cargo space to Europe. In this crisis America forgot a basic lesson in war transportation which had been learned in the Civil War. Back in 1864, in his Atlanta campaign, General Sherman had insisted that the 160 freight cars that arrived daily had to be unloaded at once and returned north for another load. In 1917 far too many cars were being used for storage purposes. In the fall of 1917, when the national shortage of freight cars was reported at 158,000, some 180,000 loaded cars were piled up at or near eastern ports with no place to go.

The railroads tried to meet the challenge of war. Immediately after war was declared, Daniel Willard (1861–1942), president of the Baltimore & Ohio Railroad and chairman and transportation expert of the Advisory Commission of the Council of National Defense, asked the nation's top rail executives to assemble in Washington, D.C. There, on April 11, 1917, nearly seven hundred railroad presidents signed a resolution in which they agreed to contribute to the war effort by running their several lines as though they constituted "a continental railway system." To coordinate their efforts the railroads created a five-man Railroads' War Board with the scholarly president of the Southern Railway, Fairfax Harrison (1869–1938), as chairman. Under Harrison's aggressive direction the group organized car pools to relieve the car shortage, urged heavier loading of freight cars, and encouraged economy in operation by suggesting the elimination of duplicate passenger trains.

Railroad executives found it difficult to operate their collective lines as a "continental" system, since any real unification was legally impossible under the Antitrust Act of 1890. Furthermore, each carrier was prone to keep and use any strategic traffic advantage it might possess. Throughout 1917 the morale of many rail employees was not high, and most lines were plagued by a constant labor turnover caused by the drafting of many railroad men and the fact that other employees were attracted by the high wages available in war plants. Shortages of labor, equipment, and capital, plus the early arrival of winter weather, brought the railroads into a fresh crisis by November 1917. On November 1 the car shortage stood at 158,000 cars and was increasing daily. In a special report to Congress on December 1, the Interstate Commerce Commission recommended that President Wilson assume control of and operate the nation's railroads. As a merely advisory body, the Railroads' War Board had been proven incapable of solving an almost impossible problem.

The government had entered the war transportation picture months before it took possession of the railroads in December

1917. A portion of the Army Appropriation Act of August 29, 1916, known as the Federal Possession and Control Act, gave the President the power to take possession of any system of transportation in wartime, should the emergency require such action. Several weeks after the declaration of war the Esch Car Service Act of May 29, 1917, gave the Interstate Commerce Commission full authority to establish rules covering the movement, distribution, exchange, and return of all railroad cars. The Commission was slow to exercise any real authority under this act.

Early in the war, under the authority of the Federal Possession and Control Act, officials of the United States Shipping Board, the Army, and the Navy all began to issue priority orders for rail shipments. Soon thousands of government officials throughout the nation were using bundles of preference tags. When placed on a freight car, the tag gave the car the right of way over untagged shipments. The tags were issued and used indiscriminately, and soon the bulk of all freight seemed to be moving under priority orders. Late in 1917 the Pennsylvania Railroad estimated that 85 per cent of the freight on its Pittsburgh division was labeled as priority traffic. Thousands of cars carrying building materials for the Hog Island shipyard were rushed to the area under priority order weeks before any unloading facilities were available. The government endeavored to bring some order to the priority confusion with the passage of a Priority Law on August 10, 1917. Robert S. Lovett (1860–1932), chairman of the executive committee of the Union Pacific Railroad, was appointed Director of Priority by President Wilson. Lovett apparently felt that he had no authority to cancel existing orders, but at least he issued very few additional ones.

By Christmas, 1917, the time for further government action had clearly come. As an early and unusually severe winter set in, the resulting drop in rail efficiency was shown by the decline in average daily freight car mileage from twenty-six miles in November to twenty-one miles in December. Following the recommendation of the Interstate Commerce Commission and acting under the authority of the Federal Possession and Control Act, President Wilson, on December 26, 1917, issued a proclamation providing for the government operation of the railroads, effective at noon December 28. In his proclamation and the accompanying statement, the President pointed out that the financial needs of the railroads and of the government could be satisfactorily met only if they were under a common direction. He also claimed that the full mobilization of the nation's economic resources required that a single authority run the railroads of the country. In his message to Congress on the subject on January 4, 1918, he paid tribute to the patriotic zeal and ability of the nation's rail executives and said: "If I have taken the task out of their hands, it has not been because of any dereliction or failure on their part, but only because there were some things which the Government can do and private management cannot."

Wilson promised full protection of the property rights of railroad owners, but it was upon his recommendation that Congress passed the Railroad Control Act on March 21, 1918, which provided that the annual compensation to the several lines should be no more than the average net operating income for the three years ending June 30, 1917. Most owners of railroad securities would have preferred a rental based upon the last year before the nation went to war, since the three-year base period included the very lean earnings of 1915. The legislation of March 1918 also provided for maintenance, repairs, and renewals by the government, with the property to be returned "in substantially as good repair and in substantially as complete equipment as it was at the beginning of the Federal control." The expenses of federal control and operation were to be met with a revolving fund of $500,000,000 created for that purpose. The law also provided that the lines were to be returned to their owners within twenty-one months following the ratification of the treaty of peace.

In his proclamation of December 26, Wilson appointed William G. McAdoo (1863–1941) Director General of Railroads. Already Secretary of the Treasury, McAdoo was chosen by Wilson because of this position, since the control of the railroads was a matter of finance as well as actual operation. As a young lawyer in New York City early in the century, McAdoo had become interested in completing railroad tunnels under the Hudson River. The financial and engineering success of the venture was matched by McAdoo's skill in public relations when in 1909 he coined the phrase "The Public Be Pleased" for his concern, the Hudson and Manhattan Railroad Company. McAdoo brought to his new position a general knowledge of railroads, years of financial experience, and proved administrative ability.

As a general rule, the new Director General allowed the operating organizations of the individual roads to remain intact. Some of the railroad presidents kept their positions, but many were replaced during the course of the war by federal managers. Final control of the roads, of course, lay in Washington, where McAdoo set up a central administrative system which was intended to weld the many lines of the country into a single great system. McAdoo naturally staffed his organization with railroad men. Walker D. Hines (1870–1934), chairman of the board of the Atchison, Topeka and Santa Fe, was appointed Assistant Director General. Early in 1918, McAdoo divided the railroads of the nation into three regions: the Eastern (east of Chicago and the Mississippi and north of the Ohio and Potomac rivers), the Western (west of Chicago and the Mississippi River), and the Southern (east of the Mississippi and south of the Ohio and Potomac rivers). To be regional director of the Eastern Region, where the worst rail congestion was located, McAdoo appointed his friend, Alfred H. Smith (1863–1924), president of the New York Central.

McAdoo and the Railroad Administration soon made a number of changes in rail operations in the interests of efficiency and a more co-ordinated service. In an effort to discourage unnecessary civilian travel, duplicate passenger-train service was eliminated, sleeping-car service was curtailed, and limited trains were obliged to handle local traffic. Trains with very few passengers were discontinued entirely. The curtailment of competing duplicate passenger service in such overserved middle western cities as Chicago, St. Louis, and St. Paul was quite justified. The annual savings in passenger traffic for the nation was estimated by the Director General to be more than 67,000,000 passenger-train miles. Despite this reduction in service, the railroads had 8 per cent more passenger traffic in 1918 than in 1917. The consolidation of timetables and ticket offices was also a useful economy. Savings in both passenger and freight service were made as the joint use of terminals, repair shops, and other equipment was initiated.

Stringent controls were introduced for railroad freight. The centralized management of the routing and distribution of traffic reduced the congestion on crowded lines and resulted in a greater use of routes that had been operating well below their full capacity. An effort was made to control the freight at the source, and terminal managers at all important points endeavored to have shipments accepted only when reports indicated that prompt delivery at the destination was possible. McAdoo vigorously applied the principle that freight should, if possible, always be sent by the shortest route. Savings were made as the Railroad Administration introduced standardization in repair work and a greater uniformity in new equipment. McAdoo ordered 100,000 new freight cars and 1,930 locomotives, all built to standard specifications. The new equipment, costing $380,000,000 and ultimately charged to the railroads, used up a major portion of the revolving fund provided by Congress. But serious freight-car shortages were a thing of the past by the late spring of 1918. The total freight moved in 1918 (440,000,000,000 ton-miles) was only 2 per cent greater than that moved under private management in 1917 (430,000,000,000 ton-miles), but the sense of crisis and confusion was notably less.

One of the first major problems facing McAdoo was that of railroad labor. Well aware of increased living costs and well-paying war-industry jobs, many rail workers were restive in 1917. On December 1, 1917, two of the operating brotherhoods made wage demands calling for nearly a 40 per cent increase in pay, and the remaining brotherhoods were considering comparable requests. Urged on by Alfred H. Smith, a man who knew, by experience, railroad work from track hand to the top executive post of the New York Central, McAdoo decided to raise wages. On January 18, 1918, he appointed a Railroad Wage Commission to investigate wages and working conditions, with any recommended wage increases to be retroactive to January 1, 1918.

The Wage Commission, which made its report on April 30, found, among other things, that the cost of living had gone up 40 per cent between December 1915 and December 1917 and that 51 per cent of all rail employees in December 1917 were making $75 per month or less. In recommending wage increases, the Commission decided to apply them to wage rates as they stood in December 1915 and to grant increases on a sliding scale which favored the lower-paid workers. Employees making up to $85 per month received at least a 40 per cent boost, the $100-a-month man went up 31 per cent, the $150 wage climbed 16 per cent, and a monthly wage of $250 received no increase at all. The great bulk of the 1,800,000 workers received substantial increases, but many of the higher-paid workers were unhappy because the new rates did violence to long-established wage differentials in the industry. The yearly cost of labor rose from $1,782,000,000 in 1917 to $2,665,000,000 in 1918. During 1918 the Railroad Administration standardized wages throughout the country and adopted the eight-hour day as normal for all types of railroad work.

Wage increases were also made after the war, and the annual average compensation per employee rose from $1,003 in 1917 to $1,419 in 1918; $1,485 in 1919; and $1,820 in 1920, the year the owners regained their lines. The portion of the railroad revenue dollar going to labor rose from 40 cents in 1917 to 55 cents in 1920. Naturally, railroad executives were bitter about the high labor costs they inherited from the years of government control. Their bitterness seems rather reasonable when one recalls that the average railroad worker's annual compensation of roughly $1,000 in 1917 was 27 per cent above the average for workers engaged in manufacturing that year. Average railroad wages in 1920 were 33 per cent above those in manufacturing.

Railroad expenses other than labor also rose rapidly during the war years. The total cost of coal and locomotive fuel nearly doubled between 1916 and 1918, although the total freight movement in the period increased only 11 per cent. As a result of the heavy increases in operating expenses, McAdoo, in May 1918, announced an average 28 per cent increase in freight rates and an 18 per cent increase in passenger fares, effective June 25, 1918. Since the boost in rates (unlike the wage increases) could not be retroactive, the increases in operating costs during 1918 were substantially higher than the increases in total revenue. The situation was not improved in the post-Armistice period of government operation. The railroad operating ratio, which had been an excellent 65.5 per cent in 1916 and an acceptable 70.4 per cent in 1917, ran to 81.5 per cent in 1918; 85.5 per cent in 1919; and 94.3 per cent in 1920. While from the standpoint of operating results government management was satisfactory, it was clearly otherwise from a financial point of view.

Shortly after the Armistice, McAdoo resigned from his many government posts and retired to private life, giving as one of his reasons the depletion of his personal resources during his years of government service. Wishing to help the man who had done so much for them, the railroad workers of the country started a campaign to have each worker contribute a dollar a year to keep McAdoo at the head of the Railroad Administration. McAdoo declined with thanks and left the government. His successor as Director General was his second in command, Walter D. Hines. Mr. Hines had the thankless task of running the lines after the war crisis was over. Faced with an unprecedented decline in rail traffic and with record expenses, he decided to save the national economy from the burden of higher rates, which the rail industry clearly deserved. Naturally, the deficits grew and grew. As admitted by Hines in April 1920, the excess of total operating expenses and rentals (paid to the several railroads) over total revenues for the twenty-six months (January 1, 1918–March 1, 1920) of federal control amounted to just over $900,000,000. This figure does not include the $204,000,000 paid the railroads in settlement of their claims for undermaintenance during federal control. Government management of the nation's railroads had been a costly venture.

The financial results of federal wartime operation of the nation's railroads are certainly not conclusive proof of the success or failure of government in business. The railroads were not taken over in 1917 for the purpose of making money. Granting that the necessity for federal operation was partially the result of governmental mistakes in both regulation and the issuance of priorities, the transportation crisis as of Christmas, 1917, still called for the action taken by Wilson. Both war freight and troops had to be moved, and the Railroad Administration made a significant contribution to the successful prosecution of the war and the final victory which followed. With victory complete, the President, in his annual message to Congress in December 1918, urged that body to give speedy consideration to the railroad problem.

The retiring Director General, Mr. McAdoo, advised that Congress provide for five years of additional federal control. Still more extreme was the Plumb Plan, named for Glenn E. Plumb (1866–1922), legal counsel for the four big railroad brotherhoods. Plumb urged that the government buy the railroads and entrust their direction to a fifteen-man board representing equally the government, railroad labor, and the operating officials. Railroad workers, the American Federation of Labor, and other labor groups all endorsed the Plumb Plan, but it received scant consideration from the general public, which wanted to see an early return to private management and direction. Congress wrestled with the railroad problem from the summer of 1919 well into the following winter. It completed its work late in February 1920, just a few days before the railroads were returned to private control on March 1.

The Esch-Cummins Act, or Transportation Act of 1920, in providing for the return of the railroads to private management, greatly increased the power and scope of the original Interstate Commerce Act. Not only was the Interstate Commerce Commission increased in size from nine to eleven members, it was also given many new powers and functions. In its supervision of railroad rates the Commission could establish minimum as well as maximum rates, and in its rate-making it was to fix them at a level which would assure a fair rate of return (originally set at 5.5 per cent) on the railroad investment. One-half of any excess income beyond 6 per cent was subject to recovery by the government. But the immediate concern was to get the railroads operating in the black, and for the first six months of private operation (March 1, 1920, to September 1, 1920) the Transportation Act guaranteed the individual lines a net railway operating income equal to half the annual rental paid during the period of federal control. This six-month guaranty cost the federal treasury $530,000,000. Higher railroad rates, long overdue, went into effect at the end of the six-month period. The freight-rate increase ranged from 25 per cent in the South and West to 40 per cent in the East.

Among other new powers granted the Commission was the supervision of all railroad-security issues and the requirement that mergers or railroad consolidations be subject to Commission approval. A marked change from earlier attitudes of regulation was the provision in the Transportation Act which directed the Commission to take steps looking toward the consolidation of the country's railroads into a limited number of systems. To comply with the law, the Commission employed Professor William Z. Ripley (1867–1941) of Harvard University, author of many standard works on railways, to formulate a complete consolidation plan for the nation. Professor Ripley suggested nineteen great systems, one of which was the merger of the Burlington with the Northern Pacific, a combination expressly denied by the Supreme Court seventeen years earlier in the Northern Securities case. The consolidation plans obviously were not compulsory, and no significant mergers based upon them took place in the generation following the passage of the Transportation Act in 1920. The ICC was also granted extensive new powers over the construction and abandonment of railroad mileage.

The legislation of 1920 created the nine-man Railroad Labor Board, three members representing the railroads, three the employees, and three the general public. The new board was given extensive jurisdiction over railroad labor disputes concerning grievances, rules, working conditions, and wages. It was hoped that the force of public opinion could be relied upon to enforce the decision of the new agency. The end of the war found railroad labor again restless, several unauthorized strikes having plagued the last weeks of government control under Director General Hines. The Railroad Labor Board entered the disputes in the spring of 1920 and on July 30, 1920, ordered a general advance averaging 22 per cent in wages for all classes of railroad labor, effective May 1, 1920. The annual cost of the increase was estimated at more than $600,000,000.

Railroad managers were not happy with the general situation as they resumed private direction of the roads in 1920. Railroad workers seemed disgruntled even though their wages had just been raised, general operating costs were high, and the year's operating ratio was a record high of 94 per cent. Strangely, the business downturn in 1921 did not cause the situation to deteriorate further. Operating revenues declined more than 10 per cent between 1920 and 1921, but operating expenses were reduced even more. Railroad employment was cut back more than 15 per cent during 1921, and wage reductions averaging about 12 per cent became effective July 1, 1921. As a result, the operating ratio declined to 82 per cent in 1921 and averaged below 75 per cent for the rest of the decade.

The return of relative railroad prosperity in the twenties did not make railroad management complacent about its recent experience with government operation. The troublesome strike of railway shopmen in the summer of 1922, serious freight-car shortages in the same year, and operating problems caused by the unusually severe winter of 1922–23 helped convince railway executives that they should embark upon a program of increasing operating efficiency. Even earlier in 1920, one result of the wartime experience had been the creation of the Car Service Division by a predecessor of the Association of American Railroads. The railroads collectively conferred upon the new agency the power to marshall and direct railroad cars to meet any emergency. Another step in cooperation and increased efficiency came in 1923 with the creation in Minneapolis of the Northwest Shippers Advisory Board, an organization of shippers and customers of the railroad seeking to make railroad transport more productive. Within a few years the nation was served by thirteen such regional boards. Each board normally met four times a year in an effort to estimate the potential needs for railroad cars and service for the next three months. Both the efficient use of rail equipment and the avoidance of car shortages were greatly aided by the activities of the Car Service Division and the Shippers Regional Advisory Boards.

Rail executives wanted even more direct action as they sought an improvement in operating efficiency. In April 1923, many of them met in New York City and adopted a comprehensive program of railroad improvement and increased efficiency. In a multi-pronged approach to their problem they agreed to: (1) increase materially their capital expenditures for additional motive power, rolling stock, and trackage facilities; (2) reduce the percentage of "bad order" equipment awaiting repair; (3) plan an earlier seasonal movement of coal; (4) plan maintenance and construction work as early in the year as possible; (5) urge more prompt loading and unloading of cars by shippers; (6) better their utilization of existing cars by increasing average car loading to thirty tons and average daily car movement to thirty miles; and (7) avoid car shortages through increased co-operation between railroads and with the Car Service Division.

Between 1923 and 1930 the railroads spent $6,741,000,000 in a capital improvement program which included $765,000,000 for new motive power, $1,760,000,000 for freight cars, $398,000,000 for passenger equipment, and $1,360,000,000 for new and improved track and roadway. As a result of these heavy purchases, more than a quarter of their locomotives and nearly 40 per cent of their freight cars were less than eight years old as of 1930. The depression thirties greatly slowed these capital expenditures, but the drive for operating efficiency continued throughout the decade. The increases in operating efficiency achieved between 1921 and 1940 can be seen in table 7.1.

The substantial increases in operating efficiency achieved between World War I and World War II were augmented by improvements in track and by advances in signaling. Back in 1916, when an all-time high of 254,000 miles of main line had been reached, there had been 143,000 miles of other track, sidings, and yard track, or an average of 56 miles of additional track for each 100 miles of main line. While the main-line mileage slowly declined in the twenties, the mileage of additional trackage increased. By 1930 there was an average of 72 miles of additional track for each 100 miles of main line. By 1941 the miles of main line had

TABLE 7.1

INCREASES IN OPERATING EFFICIENCY, RAILROAD FREIGHT SERVICE

	1921	1940	Increase (per cent)
Average freight car capacity, in tons	42.5	50.0	17
Daily mileage per serviceable freight car, in miles	25.8	38.7	50
Daily ton-mileage per serviceable freight car, in ton-miles	448	661	48
Length of average freight train, in cars	37.4	49.7	33
Net tonnage carried by average freight train, in tons	651	849	30
Average freight train speed (including all stops), in m.p.h.	11.5	16.7	45
Net ton-mileage per freight train hour, in ton-miles	7,506	14,027	87
Annual ton-mileage of freight service per railroad employee, in ton-miles ...	181,000	358,000	98
Pounds of coal required to move 1,000 gross tons one mile	162	112	31 (savings)
Payments for loss and damage of freight, per revenue car loaded	$2.35	$.55	77 (savings)

dropped to 231,971, but the total trackage was 403,625 miles, or an average of 74 miles of additional trackage for each 100 miles of main track. The weight of the rail had also been increased during the years of peace from an average weight of 83 pounds per yard in 1921, to 96 pounds per yard in 1941. In the latter year well over 40 per cent of all main-line rail weighed more than 100 pounds to the yard.

Comparable advances came in signaling during the interwar years. In 1921 only 39,000 miles of line, or 16 per cent of the total, were equipped with automatic block signals. Twenty years later this had increased to 66,000 miles, or 28 per cent of the main-track mileage. Even more important was the start made in the late twenties and thirties in Centralized Traffic Control. Centralized

Traffic Control is the operation of trains over a division, or a portion thereof, in which a single operator-dispatcher at a control and indicating machine sets switches and trackside signals for the movement of all trains. Since the dispatcher can arrange for the meeting of trains in the most efficient way, savings in time are great. There were 36 installations of Centralized Traffic Control on 341 miles of line in the country in 1929. This had grown by 1941 to 229 installations serving 2,163 miles of road. The reduction of congestion on heavily used lines with the new type of control soon proved phenomenal. During the war years Centralized Traffic Control was added to much southern and western single track mileage. By 1945, when the trackage thus controlled had tripled to 6,495 miles in 328 installations, Centralized Traffic Control had made a major contribution to the successful railroad war effort.

The effect of World War II on the nation's railroads was felt long before Pearl Harbor. The national-defense program and the requirements of the Lend-Lease Act brought a return of prosperity to the railroads after the long years of depression. The year 1940 found the railroads of the country with higher total operating revenues than for any year since 1930 and with an operating ratio (72 per cent) that was lower than for any year since 1929. Business was even better the following year, when operating revenues climbed 25 per cent. In 1941 freight ton-mileage reached 475,000,000,000, a figure 17 per cent higher than the 1918 World War I peak and 6 per cent above the previous all-time-record year of 1929. The operating ratio dropped to 69 per cent, the lowest since 1916, and the rate of return on the net property investment rose to 4.28 per cent, the highest since 1929. After years of red ink, receivership, and low income, the new rail prosperity was most welcome.

As America faced a world war for the second time in a generation, the nation's railroads made every effort to escape the government control they had endured from 1917 to 1920. But World War I had quite clearly shown the need for at least some form of centralized direction over the transportation of a nation at war. President Franklin D. Roosevelt (1882–1945) took the first step toward a co-ordination of wartime transport in May, 1940, when he appointed Ralph Budd (1879–1962) Transportation Commissioner of the Advisory Commission to the Council of National Defense. Mr. Budd, who as president of the Burlington Railroad had revitalized passenger traffic on his line in 1934 with the streamlined diesel "Zephyr," was concerned with truck, bus, air, and pipeline transportation as well as railroads. In January 1941, his duties were shifted to a new agency as he was appointed Director of the Transportation Division in the Office of Emergency Management. Though he had no mandatory powers, Budd endeavored, through co-operation and consultation with the railroads, the ICC, and other advisory groups, to avoid the troubles that had plagued transport in World War I.

After Pearl Harbor a more comprehensive supervision of transportation seemed to be needed, and on December 18, 1941, Roosevelt set up the Office of Defense Transportation (ODT), appointing Joseph B. Eastman (1882–1944), veteran member of the ICC and Federal Co-ordinator of Transportation from 1933 to 1936, Director. The ODT was responsible for co-ordinating all transportation facilities, estimating war traffic requirements, and directing domestic war traffic through co-operation with ocean shipping so that port congestion could be avoided. In short, the ODT, through full co-ordination and co-operation with the total transport facilities of the nation, was trying to avoid the mistakes and confusion that had occurred in 1917 and 1918. The railroad industry was most anxious to substitute co-operation for compulsion. Individual railroads, the Association of American Railroads, along with its Car Service Division, and the Shippers Advisory Boards all co-operated in the fullest possible way. As a result, the government did not find it necessary in World War II to manage the nation's railroads except for a very brief period of labor difficulty in December 1943 and January 1944.

Complete co-operation among the railroads was essential as they faced the demands of war in December 1941. The depression thirties plus the competition of rival transport agencies since World War I had forced a substantial decline in railroad facilities. As compared to the end of 1916, American railroads at the end of 1941 had 25 per cent fewer freight cars, 30 per cent fewer passenger cars, and 32 per cent fewer locomotives. The number of rail employees in 1941 was down nearly a third from 1916. However, average freight-car capacity and average locomotive tractive effort had grown significantly in the quarter-century. Thus the 1,703,000 freight cars in 1941 had a total carrying capacity (85,682,000 tons) only 7 per cent below those of 1916, and the 41,771 locomotives actually had an aggregate tractive effort slightly above that available in the earlier year.

With a total carrying capacity certainly no larger than that of World War I, the nation's railroads in World War II moved a total traffic of much greater volume. The total ton-mileage of freight carried in each of the four war years (1942 through 1945) was more than 50 percent greater than in 1918, and in the peak year, 1944, the 737,000,000,000 ton-miles was 82 per cent above the top year of the earlier war. Passenger traffic also quickly jumped. Total passenger-mileage in 1942 was a quarter larger than that of 1918, and each of the years 1943, 1944, and 1945 saw a passenger traffic that more than doubled the 1918 figure. When one remembers the low base of depression rail operations, the increase was even greater. Rail freight in 1944 more than tripled the figures for 1932, the low year of the depression, and passenger traffic climbed almost sixfold in the same years.

Between Pearl Harbor and the end of the war, American railroads furnished 97 per cent of all domestic troop movements and about 90 per cent of all domestic movements of Army and Navy equipment and supplies. In the forty-five months of war, 113,891 special troop trains moved 43,700,000 members of the armed forces, or an average of nearly a million men a month. This was more than twice the average monthly troop movement in World War I. Because of shortages of rubber and gasoline and because German submarines drove coastal tankers from the open sea, most of the increased wartime traffic was carried by rail. Between 1941 and 1944 the railroads handled 77 per cent of the new intercity passenger traffic and 83 per cent of the new commercial freight business. The boom in new traffic was most noticeable in the movement of oil to the eastern seaboard. In the summer of 1941 rail deliveries of oil to the East amounted to only 11,250 barrels daily, but as German submarines pinched off coastal tankers, rail deliveries to the East rose to more than 1,000,000 barrels a day. The flood of new rail business was taken care of chiefly with equipment in use at the time of Pearl Harbor. During the war years the War Production Board was so stingy with material allocations to the railroads that passenger cars were increased in number by only 1 per cent, the number of freight cars increased only 3 per cent, and the number of locomotives grew only 4 per cent.

The successful record of American railroads during World War II was made possible chiefly by the tremendous increase in operating efficiency achieved in the two decades between the two world conflicts. This new efficiency was so marked that whereas daily average shortages of freight equipment had been 113,000 cars in 1917 and 52,000 cars in 1918, in World War II they dropped to a daily average of only 10,000 cars in the worst year, 1945. The record achieved by the railroads was also made possible by differences in the nature of the traffic in the two wars. In World War I, a one-front war, most traffic went to eastern ports, subsequently requiring a large amount of empty-car mileage westward. A more efficient use of equipment was possible in the two-front

Second World War, since there were large shipments headed west as well as east.

This heavy traffic across the vast distances of the mountain West clearly contributed to a second transport difference in the two conflicts. In World War II the average haul, both for freight and passengers, was much greater than in World War I. In 1918 the average freight haul was 321 miles, while in 1944 it was half again as large, or 494 miles. The total of originated revenue freight in 1944 was 1,491,000,000 tons, or only 18 per cent larger than the 1,263,000,000 tons originated in 1918. Since the cars were smaller in the first war the 44,592,000 cars of revenue freight loaded in 1918 actually exceeded by nearly 3 per cent the 43,408,000 cars loaded in 1944. In the passenger traffic of World War II the increase in average length of trip was even greater. Whereas in 1918 the average passenger trip had been 39 miles, it was 105 miles in 1944. It was the longer average trip, much more than any increase in passengers, which explained the 1944 record-high passenger business of 95,549,000,000 passenger-miles, a figure that was over twice the total of 1918. The number of revenue passengers carried in 1918 was 1,085,000,000, a figure actually 19 per cent higher than the 910,000,000 carried in 1944. The longer haul and longer trip, typical of wartime business, were major keys in explaining the success of rail traffic in the early forties. Not only did they make for great increases in operating efficiency, but the substantially higher revenue they brought in assured the rail lines of a total operation that was increasingly profitable.

As was World War I, the Second World War was a time of trouble and difficulty for railroad labor. With the new bulk of wartime traffic, the ranks of railroad labor naturally grew. Despite a loss of 351,000 men to military service, the number of rail employees expanded nearly a third during the war, growing from 1,140,000 in 1941 to 1,420,000 in 1945. In 1941 there was strong agitation from all types of railroad workers for wage boosts and for vacations with pay. When the recommendations of an emergency mediation board headed by Wayne L. Morse (1900–1974), University of Oregon Law Dean and later United States Senator, proved unsatisfactory to the workers, a strike was called for Sunday, December 7, 1941. President Roosevelt's intervention brought additional wage concessions, and the strike was averted with an agreement announced on December 2, 1941. Operating employees received an average wage increase of 7.5 per cent and the lower-paid non-operating workers received an average boost of 13.5 per cent.

The continued climb in the cost of living, wage increases in industry, and the obvious prosperity of the railroads soon made the workers anxious for a second increase. When their wage demands were not met, the operating unions called a strike for December 30, 1943. As they had done in 1941, the workers were able to use their political influence to obtain the aid of President Roosevelt in getting increases above the recommendations of the mediation board. Needing time to work out a satisfactory agreement, Roosevelt took over the railroads. The War Department had possession of the lines from December 27, 1943, until January 18, 1944, but operations remained in the hands of railroad management and government control was nominal. As a result of the wage increases, the average annual compensation for rail workers rose from $2,045 in 1941 to $2,720 in 1945, a gain of 33 per cent. Consumer prices in the same years had climbed 19 per cent. The prospering railroads paid higher wages but still had an average labor cost per revenue dollar of only 39 cents for the five years from 1941 through 1945. Railroad rates were not raised substantially during the war, and the average revenue of 0.96 cents per ton-mile in 1945 was as low as during the depression thirties.

Of the 351,000 railroad employees who went off to war, more than a tenth continued to serve the transportation needs of the nation as members of railway battalions and other units of the Military Railway Service. During the war the 43,000 officers and men in the Military Railway Service operated rail mileage in such far-off places as Alaska, England, French Morocco, Algeria, Sicily, Italy, France, Germany, India, Burma, and the Philippines. The director general of the organization, Major General Carl R. Gray, Jr. (1889–1955), like many members of his staff, had been a railroad executive before the war.

One of the most favorable aspects of wartime rail traffic was the prosperity it brought to the rail industry. Business was so good that the operating ratio for the five war years (1941–45) averaged only 67.7 per cent, the lowest five-year average in the twentieth century, except for the years 1901 through 1905, and nearly twenty points below the 87 per cent average figure for the three years of government operation from 1918 to 1920. The average rate of return on the net property investment during the war years was 4.97 per cent, the highest five-year average for twenty-five years. While cash dividends paid during the war were not of record size, the industry put its financial house in order by retiring nearly $2,000,000,000 of funded debt. The industry's net funded debt in the hands of the public at the end of 1945 stood at $8,659,000,000, the lowest figure since well before World War I. In contrasting the financial well-being of the railroads under private direction in World War II with the government operation of World War I, the Association of American Railroads pointed out that while government operation had cost the American taxpayer $2,000,000 a day, the railroads in World War II (1942–45) paid average federal income taxes of more than $3,000,000 each day.

Clearly, a variety of reasons contributed to the success of the railroads in meeting the transportation needs of the Second World War. The volume of freight and passenger traffic moved in the years 1941 through 1945 seemed greater because of the low rate of traffic in the depression thirties. But the major fact was that increased operating efficiency, improved trackage, better signal-

1940s. The best efforts of the railroads to keep their business—by improving operating efficiency, introducing streamlined and air-conditioned passenger service, and building new forms of motive power—seemed to do little more than slow their decline. By 1965 they controlled no more than 18 per cent of the total intercity commercial passenger business and only 44 per cent of the inter-city freight traffic. The drop in the passenger traffic was the most marked, not only in volume, but also in profits. The railroads claimed annual deficits for their passenger service for each year since 1929, except for the four war years, 1942 through 1945. Passenger deficits brought more curtailment of service, and by 1965 passenger service was available on no more than 37 per cent of the national rail network.

Despite prospects that seemed bright at the end of World War II, the economic position of American railroads would continue a decline which had started by 1920 (see table 8.1). Railroads failed to share equally in the bulge of postwar freight traffic that came in the late forties and fifties. Although the nation had a widespread prosperity, the railroads, since 1945, had to be satisfied with a rate of return on their net property investment that ranged from 2.76 per cent to 4.31 per cent and averaged but 3.64 per cent. In 1957, when the four largest American airlines had a net income that averaged 7.75 per cent of their total invested capital, the four largest railroads had a net income of only 2.9 per cent.

The relative falling off of rail traffic in the twentieth century was most intimately associated with the appearance of new, competitive transport facilities. The year 1916 was not only the year of peak rail mileage, but it also marked the resumption of an interest by the federal government in highway construction when Congress passed the Federal Aid Road Act providing grants-in-aid to the several states. Henry Ford (1863–1947), with his durable Model T, and hundreds of other automobile manufacturers were so busy selling their product that an increased interest in good

TABLE 8.1

AMERICAN RAILROADS IN DECLINE

Year	Mileage	Total Invest-ment (billions)	Operating Revenues (millions)	Railroad Employ-ees	Average Annual Wages		Annual Freight Ton-Mileage per Employee
					(Curr. $)	(1920 $)	
1920	253,000	$20	$ 6,310	2,076,000	$ 1,820	$1,820	199,000
1933	246,000	25	3,138	991,000	1,445	2,225	253,000
1945	227,000	24	8,986	1,439,000	2,720	3,019	475,000
1965	212,000	26	10,425	655,000	7,490	4,719	1,076,000
1987	163,000	47	26,622	247,000	37,716	6,713	3,919,000

Source: John F. Stover, "Railroads," p. 909, in *The Reader's Companion To American History* (1991), with permission from Houghton Mifflin Company.

roads was inevitable. Private motor vehicle registrations reached 3,367,000 in 1916, and as the typical American family bought its "first car" in the prosperous twenties, the automobile registration figures climbed to 23,000,000 in 1929. Now that it was statistically possible for the entire population to be speeding along the new highways at one time, there was naturally a tremendous increase in non-commercial intercity traffic. Total intercity traffic, which had been but 42,000,000,000 passenger miles in 1916, climbed fivefold to 198,000,000,000 passenger miles in 1929, with more than three-quarters of the total consisting of private automobile travel.

Noting the convenience, cleanliness, and novelty of the new motorbuses available in the twenties, more and more people used them for short trips. Competition was keen among the thousands of small bus companies, but consolidation and longer through service appeared with the creation of the Greyhound system in 1929. Noting the trend, some railroads began to operate bus systems of their own. Before the end of the decade railways were running 1,000 motorcoaches over 10,000 miles of route. By 1930 the highway bus had achieved a respectable popularity and was furnishing 7,000,000,000 passenger-miles of service yearly, or 18 per cent of the commercial intercity traffic.

The advantages of extensibility, adaptability, and convenience were obvious early in the development of highway transport. The more than 3,000,000 miles of roads and highways in the nation gave the bus and truck a flexibility of service that trains could never hope to equal. Hundreds of small villages and hamlets which had never seen a railroad could be served by the new highway facilities. By the end of World War II about half the roads of the country were improved highways. The smaller size of the highway vehicle (as compared to rail facilities) also permitted a greater flexibility of scheduling than was normal for the railroad. The door-to-door convenience and the frequent schedules, especially for the short haul, soon brought expansion to the highway freight business. By the end of the twenties, the 3,000,000 trucks in the country carried less than 4 per cent of the total intercity freight ton-mileage, but the growth was great in the depression thirties. On the eve of Pearl Harbor, highway trucks were carrying 10 per cent of the intercity freight, or an annual total of 62,000,000,000 ton-miles.

Unregulated by the federal government until 1935 and aided by a right of way that was virtually free (as compared to the railroads), truck operators took much traffic away from the railroads. The ability of the furniture van to move uncrated household goods meant the nearly total loss of such traffic by the railroad. In the years since World War I the railroads lost the vast majority of their less-than-carload freight, with the total of such traffic dropping from 51,000,000 tons in 1919 to 1,400,000 tons in 1965. The railroads also lost the bulk of their freight in animals and animal products to the truck, the volume of such business declining from 35,000,000 tons in 1919 to 9,500,000 tons in 1960. Between 1922 and 1960 railroad-owned stock cars dropped in number from 80,000 to 31,000, and railroad-owned refrigerator cars dropped from 63,000 to 25,000. By the early fifties the truck was carrying major portions of the country's eggs, poultry, milk, fruits, and vegetables. Larger and larger shipments of manufactured goods, such as motor vehicles out of Detroit, were also being delivered via the highway. The average truck haul was much shorter than that of the railroad, but even this changed during the fifties. The growing trend toward heavier long-haul truck shipments was indicated in the summer of 1959 when the New York Thruway conducted tests on multi-unit "Turnpike Trains," which could have a maximum length of 98 feet and a maximum weight of 65 tons.

Competition from airlines came in these same years. Aviation had proved itself by World War I, and in May 1918, the Post Office Department started to operate airmail service between New York City and Washington, D.C. Transcontinental airmail service, including night flying, had been firmly established by 1925 when the Kelly Act authorized the Postmaster General to contract with private companies to carry the mail. As private contractors started to fly the mail, they also took a few passengers, carrying a total of 5,800 passengers in 1926. The next year, Charles A. Lindbergh (1902–74) gave the infant industry a great boost with his solo flight across the Atlantic to Paris. Soon passenger service across the nation was available on a combined rail-air route provided by the Pennsylvania Railroad, Transcontinental Air Transport, and the Santa Fe Railway. Passengers crossed the nation in forty-eight hours, going by rail at night and flying in the daylight hours. All-air service soon appeared, and 73,000,000 passenger-miles were flown in 1930.

Although expansion continued in the thirties to more than a billion passenger-miles in 1940, the total airline passenger service was less than 3 per cent of the intercity commercial passenger movement. The great growth in air travel came after World War II, and by 1957 roughly a third of all commercial passenger traffic was by air carriers. For many years the speed advantages of air travel were at least partially offset by the greater danger involved. In the forties both bus and train travel were roughly ten times as safe as air travel on a basis of passenger fatalities per miles traveled. In the fifties this difference in safety declined somewhat. In 1958, when the total passenger-mileage of the three facilities was almost equally divided among them, the total passenger fatalities were: rail, 60; airline, 114; and bus, 120. In 1959, however, the hazards of air travel were statistically, fourteen times as great as those of railroad travel.

In the twentieth century railroads met competition from certain transportation facilities that had been established some time earlier. The first oil pipeline was built in 1865 (its builders used two-inch wrought-iron pipe) from Titusville, Pennsylvania, to a rail junction five miles distant. In 1867, William H. Abbott (1819–1901), a pioneer petroleum producer, formed the Pennsylvania Transportation Company, the first great consolidation of pipelines. In the same years that John D. Rockefeller was demanding rebates from the railroads for his oil shipments, he was also building and acquiring long-distance pipelines. By the turn of the century he controlled a network of 40,000 miles. In 1916 the pipelines of the nation carried 21,000,000,000 ton-miles of liquid freight, or 4.4 per cent of the total intercity freight traffic. As the age of the automobile arrived, both the production and the shipment of oil products increased greatly. The pipeline system had grown to 150,000 miles by 1947, and by 1965 more than 17 per cent (280,000-000,000 ton-miles) of all intercity freight was moving by pipeline. Rail shipments of petroleum products (crude and refined) dropped from 24 per cent of the total oil production in 1929 to 6 per cent in 1953. The railroad loss was the greater as more and more factories shifted from coal, long a leading item of rail traffic, to oil or natural gas as fuel.

Water transport was both the oldest and the least threatening competitor of the railroads. Passenger use of the inland waterways had never been of any significant size and during most of the years since 1916 never rose above 3 per cent of the commercial intercity traffic. Because of its lower rates, water-borne freight remained important. Freight on the inland waterways of the nation ranged from 15 to 18 per cent of the total ton-mileage during the period from 1916 to 1965.

Until the forties the great bulk of domestic water freight was on the Great Lakes rather than canals or rivers. Lake traffic increased very slowly, and the ton-mileage figures for the prosperous fifties generally were no more than a quarter larger than those of the late twenties. By contrast, traffic on the rivers and canals boomed, especially since the completion of an extensive program of inland-waterway improvements. In 1955, 1956, and 1957 the annual ton-mileage of freight on the canals and rivers of the United States was more than ten times the amount for the late twenties. Canal and river freight movement was roughly equal to that on the Great Lakes.

As the railroads saw their freight and passenger traffic increasingly shift to different facilities in the generation after World War I, they became critical of the role of the government, with regard to both the public promotion and the regulation of the new facilities. A 1965 report by the Interstate Commerce Commission indicated that between 1921 and 1965 some $216,000,000,000 of public money had been provided for highways and streets. Of this sum, less than half came from highway-user taxes, and a total of $43,000,000,000 had been supplied by the federal government. Federal expenditures for rivers and harbors amounted to more than $3,500,000,000 in the years 1922 through 1953. From 1925 to 1960 the federal government spent over $8,000,000,000 for a system of airway facilities and the construction of numerous civil airports. Unlike the railroad land-grant aid of the mid–nineteenth century, where a *quid pro quo* in the form of reduced rates for government freight and troops was provided, public aid for twentieth-century transport facilities was not accompanied by a discount for government business.

As the railroads watched the government build highways for truckers, dredge channels for barge companies, and construct airports for private airlines, they became unhappy with their annual tax bill. Between 1950 and 1956 the railroads of the country paid to the federal, state, and local governments taxes that averaged $1,130,000,000 a year, or 11 per cent of their total revenues. For every dollar paid to its stockholders in dividends in those years, the railroad industry paid more than three dollars in taxes. The federal income tax took a major share, but a large portion was in the form of local property taxes. In Cut Bank, Montana, where a new airport had just been built with federal funds, the Great Northern Railway, along with all the other property owners in the locality, was taxed to maintain the new air terminal for the use of Western Airlines. In 1957, when 587 passengers embarked at the Cut Bank airport, the Great Northern Railway paid $3.82 in local taxes for each departing passenger. The local property tax on the Western Airlines amounted to four cents per passenger! In Chicago the Union Station paid taxes in 1956 of $913,000 while city-owned Midway Airport paid no taxes at all.

Feeling themselves excessively regulated, railroads also noted with disapproval the difference in degree with which some of their competitors were regulated. Only about a third of all intercity truck traffic was subject to the economic control of the Interstate Commerce Commission in 1956. The trend toward unregulated truck traffic had been growing since World War II. Even a smaller portion of domestic water-borne commerce was regulated; between 1947 and 1955 less than a seventh of such traffic was under federal regulation. With considerable logic, railroad management pointed out the difficulty of their position in comparison with their largely unregulated water and highway competitors. The rail-roads especially criticized the reluctance or refusal of the ICC to permit railroads to lower their rates as they sought to compete with unregulated carriers. In a number of instances the ICC refused to let the railroads reduce their rates, based as they were on inherently lower rail costs, and instead forced the railroads to have rates high enough to permit the traffic to go by barge or truck.

No one can claim that the railroads were unregulated since the First World War. The Transportation Act of 1920 provided a complete regulation of the nation's rail lines. As annual rail revenues dropped to $3,000,000,000 (lower than any year since 1915) in the early depression thirties, deficits appeared in the rail net income in 1932, 1933, and 1934. The Interstate Commerce Commission refused the industry's request in 1931 for a general 15 per cent hike in freight rates and instead permitted selective surcharges on certain traffic, the pooled revenue therefrom to be lent to needy railroads to permit them to meet their fixed charges.

Upon the recommendation of the ICC, the Emergency Railroad Transportation Act was passed by Congress in June 1933. This legislation established the Federal Co-ordinator of Transportation, who was supposed to promote a new economy of rail operation through the elimination of duplicated services and facilities, the general reduction of expenses, and the lowering of fixed charges wherever possible. Joseph B. Eastman, the Co-ordinator during the three-year life of the Act, achieved only a modest success in his drive for operating economy. Most carriers (who viewed many of the suggestions of the Co-ordinator as impractical) were reluctant to cooperate fully in the venture. The whole effort was rendered largely ineffectual by a provision of the original act which prohibited any economy involving a reduction in rail employment. The 1933 legislation also changed the rate-making rule of 1920, which had sought rates that would assure a fair rate of return on the value of the rail property. The new rule of 1933 simply required that the ICC fix rates that would be fair to carriers as well as shippers. In the middle thirties Congress also passed new railroad-labor legislation providing for extensive arbitration and mediation procedures. A generous railroad retirement program was established with legislation in 1934, 1935, and 1937.

Recovery was slow for the railroads. The average rate of return on the net property investment in the thirties was just over 2 per cent, and in 1938, which again showed a net income in the red, companies with 31 per cent of the total mileage of the nation were either bankrupt or in receivership. Urged on by the ICC, Congress considered new legislation, and on September 18, 1940, President Roosevelt signed the Transportation Act of 1940. In a declaration of policy, the new law stated the principle of treating all carriers alike. A start was made in this direction, since the Act provided that domestic water carriers were under the regulations of the ICC. No changes of major significance were made in railroad regulation, although minor modifications were made in the long-and-short-haul clause and the land-grant rate reductions were repealed, except for military freight and personnel. In mid–twentieth century the nation's railroads were still under nearly total regulation, even though the days of monopoly were long since gone. In the middle fifties in England, France, and Canada, where the government owned all, or many, of the rail lines, the economic regulation was much less severe. In each of these three countries the railroads had much freedom in rate-making, especially in the granting of low rates. American railroads achieved the first steps toward such freedom from regulation in the Transportation Act of 1958.

The financial advantages that might come with merger are great. Savings in the elimination of terminals, excess trackage and maintenance, improved car routing and use, and centralized purchasing and repair work would seem to be huge. *Fortune* magazine (August 1958) estimated that savings would amount to at least $1,000,000,000 yearly if the railroads of the nation were combined into four great regional systems. But some consolidations were neither economical nor in the public interest. Such a rail merger was that put together in the twenties by the brothers Mantis James Van Sweringen (1881–1935) and Oris Paxton Van Sweringen (1879–1936). Originally real estate promoters of Shaker Heights, near Cleveland, the brothers bought control of the Nickel Plate from the New York Central in 1916 for $8,500,000. In the next fifteen years, through a combination of financial juggling and an amazing system of holding companies, they managed to gain at least indirect control over a rail system of more than 21,000 miles, which included, in addition to the Nickel Plate, the Chesapeake & Ohio, the Wheeling and Lake Erie, the Kansas City Southern, the Erie, the Pere Marquette, and the Missouri Pacific. The paper empire defaulted on a loan in 1935 and quickly collapsed.

Proposals for solid and more sensible mergers appeared in the middle and late fifties. While the suggested New York Central–Pennsylvania consolidation misfired, the autumn of 1959 saw the completion of the merger of two coal roads, the Norfolk and Western, and the Virginian. A year later the ICC approved the Erie-Lackawanna merger. Other major proposed or pending mergers that year included: (1) Chesapeake & Ohio–Baltimore & Ohio; (2) Rock Island–Milwaukee; (3) Seaboard–Atlantic Coast Line; and (4) Great Northern–Northern Pacific–Burlington. Since all rail mergers required ICC approval, no consolidation effort ever moves rapidly.

One result of new competition from other transportation facilities was the abandonment of substantial railroad mileage. From an all-time high of 254,000 miles in 1916, the nation's rail network dropped to 220,000 miles by 1959, a decline of more than 13 per cent for the long generation. Most of the 34,000 miles of abandoned trackage was eliminated in the years after 1930. In 1930 the country's rail system still stood at 249,000 miles, and in that year the mileage of all track operated (including double track, sidings, etc.) was at an all-time high of 430,000 miles. Fifteen states, most of them west of the Mississippi River, actually added new rail mileage during the twenties, and in both 1928 and 1929 the national total for main line in operation increased slightly. The depression thirties quickly changed this trend. More than 15,000 miles of road were abandoned in that decade, and every state except Rhode Island (which built three miles of new line) participated in the decline. Every year from 1932 through 1943 saw at least 1,200 miles of line abandoned, and in 1935 the total was over 2,000 miles. Nearly 10,000 miles of additional first-line trackage were dropped in the forties, and the national total stood at not quite 224,000 miles in 1950. Since 1950 the rate of abandonment has definitely been slower. Both the growth and the decline of railway mileage are shown in tables 8.2 and 8.3.

Since World War I most states and sections of the nation abandoned rail mileage at near the national average of from 10 to 15 per cent. Four western states—Montana, North Dakota, Oregon, and Texas—held their rate of decline to less than 5 per cent. At the other extreme, eight states—Arkansas, Colorado, Louisiana, Maine, Michigan, Nevada, New Hampshire, and North Carolina—dropped 20 per cent or more of their 1916 rail mileage. Colorado, which dropped much narrow-gauge trackage, had the record rate of abandonment of 30 per cent. Railroads in the eastern district (east of Chicago and north of the Ohio River and North Carolina) had a rate of mileage reduction since 1916 of nearly 15 per cent, while southern and western state mileage declined at just above 12 per cent. This modest difference is perhaps explained by the generally greater prosperity of western and southern lines. In 1957 the ten major eastern roads had an average net income on invested capital of only 3.6 per cent; the ten major western and southern lines averaged a 4.8 per cent rate of return.

Even though the railroads reduced their mileage in the generation after World War I, they strove to meet the challenge of new

TABLE 8.2

GROWTH AND DECLINE OF RAILWAY MILEAGE BY STATES

	1840	1860	1880	1900	1920	1940	1959
Alabama	46	743	1,843	4,226	5,378	4,996	4,647
Arizona	...	...	349	1,512	2,478	2,228	2,166
Arkansas	...	38	859	3,360	5,052	4,482	3,950
California	...	23	2,195	5,751	8,356	7,947	7,758
Colorado	...	...	1,570	4,587	5,519	4,552	3,786
Connecticut	102	601	923	1,024	1,001	887	825
Delaware	39	127	275	347	335	295	293
Florida	...	402	518	3,299	5,212	5,218	4,670
Georgia	185	1,420	2,459	5,652	7,326	6,334	5,864
Idaho	...	...	206	1,261	2,877	2,746	2,685
Illinois	...	2,790	7,851	11,003	12,188	11,949	11,211
Indiana	...	2,163	4,373	6,471	7,426	6,889	6,601
Iowa	...	655	5,400	9,185	9,808	8,950	8,569
Kansas	...	...	3,400	8,719	9,388	8,564	8,239
Kentucky	28	534	1,530	3,060	3,929	3,691	3,543
Louisiana	40	335	652	2,824	5,223	4,357	3,938
Maine	11	472	1,005	1,915	2,295	1,882	1,786
Maryland (and D. C.)	213	386	1,040	1,408	1,472	1,402	1,176
Massachusetts	301	1,264	1,915	2,119	2,106	1,793	1,654
Michigan	59	779	3,938	8,195	8,734	7,303	6,665
Minnesota	...	...	3,151	6,943	9,114	8,421	8,205
Mississippi	...	862	1,127	2,920	4,369	3,919	3,659
Missouri	...	817	3,965	6,875	8,117	7,042	6,595
Montana	...	...	106	3,010	5,072	5,149	4,969
Nebraska	...	...	1,953	5,685	6,166	6,044	5,721
Nevada	...	...	739	909	2,160	1,941	1,647
New Hampshire	53	661	1,015	1,239	1,252	1,002	869
New Jersey	186	560	1,684	2,259	2,352	2,108	1,915
New Mexico	...	...	758	1,753	2,972	2,812	2,473
New York	374	2,682	5,957	8,121	8,390	7,739	6,541
North Carolina	53	937	1,486	3,831	5,522	4,668	4,310
North Dakota	...	...	...*	2,731	5,311	5,266	5,255
Ohio	30	2,946	5,792	8,807	9,002	8,501	8,334
Oklahoma	...	...	289	2,151	6,572	6,302	5,802
Oregon	...	...	508	1,724	3,305	3,385	3,149
Pennsylvania	754	2,598	6,191	10,331	11,551	10,328	9,126
Rhode Island	50	108	210	212	211	194	181
South Carolina	137	973	1,427	2,818	3,814	3,466	3,282
South Dakota	...	...	1,225*	2,850	4,276	4,006	3,920

TABLE 8.2

(*continued*)

	1840	1860	1880	1900	1920	1940	1959
Tennessee	...	1,253	1,843	3,137	4,078	3,573	3,402
Texas	...	307	3,244	9,886	16,125	16,356	14,769
Utah	...	...	842	1,547	2,161	2,082	1,734
Vermont	...	554	914	1,012	1,077	919	811
Virginia	147	1,379	1,893	3,779	4,703	4,261	4,135
Washington	...	...	289	2,914	5,587	5,243	4,979
West Virginia	...	352	691	2,228	3,996	3,831	3,680
Wisconsin	...	905	3,155	6,531	7,554	6,639	6,194
Wyoming	...	...	512	1,229	1,931	2,008	1,882

*North and South Dakota were combined in Dakota Territory prior to 1890.

TABLE 8.3

GROWTH AND DECLINE OF RAILWAY MILEAGE IN THE UNITED STATES

1830 23	1880 93,267	1920 252,845
1840 2,808	1890 163,597	1930 249,052
1850 9,021	1900 193,346	1940 233,670
1860 30,626	1910 240,439	1950 223,779
1870 52,922	1916 254,037	1959 217,565

competition with increased operating efficiency. Unfriendly critics of the railroads in recent years have often viewed the industry's management as being composed of old fogies, averse to change and still living in the nineteenth century. Actually, the twentieth century has been one continuous period of increased productivity. Recent studies indicate that the average annual rate of increased productivity per rail worker has approximated 2 per cent since 1889. Table 8.4 indicates the increased railroad efficiency from 1921 to 1960.

As was noted earlier in chapter 7, the growth of rail operating efficiency was marked in the years between the two world wars (1921–40). While the trend toward greater productivity varies in

TABLE 8.4

INCREASE IN OPERATING EFFICIENCY, RAILROAD FREIGHT SERVICE

	1921	1940	Increase over 1921 (per cent)	1960	Increase over 1940 (per cent)
Average freight-car capacity (in tons)	42.5	50.0	17	55.4	11
Daily mileage per serviceable freight car (in miles)	25.8	38.7	50	45.6	18
Daily ton-mileage per serviceable freight car (in ton-miles)	448	661	48	953	44
Length of average freight train (in cars)	37.4	49.7	33	70.2	41
Net tonnage carried by average freight train (in tons)	651	849	30	1,466	73
Average freight train speed (including all stops; in m.p.h.)	11.5	16.7	45	19.5	17
Net ton-mileage per freight train hour (in ton-miles)......................	7,506	14,027	87	28,587	103
Annual ton-mileage of freight service per employee (in ton-miles)............................	181,000	358,000	98	733,000	104

detail, there was no material decline in the rate of increase in the forties and fifties. Both of the periods (1921 to 1940 and 1940 to 1960) reviewed in table 8.4 show greater increase in transport productivity than the industry experienced in the two decades preceding World War I.

Many gains in efficiency and operational progress came as the railroads individually and collectively engaged in systematic research projects. While many of the larger companies set up their own programs, much research activity was carried on by the two hundred standing committees of the Association of American Railroads or at the Association's own railroad laboratory and research center on the campus of the Illinois Institute of Technology in Chicago. The attention of the research staffs was directed toward such varied and diverse subjects as continuous welded rail, fatigue strength of structural welds, grade-crossing protection, roller bearings, use of alloys for lightweight car construction, safe transportation of explosives, simplification of tariffs, and improvement of accounting procedures.

Much was done in the area of railroad communications from 1914–1960. Radio, first used experimentally by the Lackawanna in 1914 and by the Nashville, Chattanooga, and St. Louis in 1922, had progressed so far by mid-century that 190 railroad and terminal companies held FCC licenses for radio stations. Since World War II more than 6,000 locomotives, 3,000 cabooses, and 1,200 offices were equipped with radio for yard or road train usage, and hundreds of walkie-talkies were distributed. Industrial television, first used by the Pennsylvania Railroad in 1954 for switching mail cars at Pittsburgh, is also available for the identification of incoming freight cars, for use in guarding rail crossings, and for the general supervision of terminal operations. Referring to its usefulness, one railroader remarked: "TV watches without coffee breaks or overtime pay and doesn't complain about rain or snow."

New operating methods and devices, the result of railroad research, have been introduced in railroad yards. Many of the yard improvement programs started shortly after World War I. In the twenties, the Illinois Central completed the Markham Yard near Chicago and named it for Charles H. Markham (1861–1930), long-time president of the line. This classification facility, when completed, was 3.5 miles long and had 113 miles of track with a capacity of 9,000 freight cars. Much of Daniel Willard's $145,000,000 improvement program on his Baltimore & Ohio in the years from 1910 to 1927 was spent on new shop and yard facilities. The Pennsylvania Railroad's giant new freight yard, built after World War II near Conway, Pennsylvania, cost $34,000,000 but was scheduled to save $11,000,000 annually in operating expenses as well as to speed up freight shipments. The typical new freight yard in the fifties was using electronic devices to identify incoming trains, locate broken wheels, retard cars, and weigh them while they were still in motion.

Faster switching and classification methods contributed to the speed-up of freight operations. Although average freight-train speed has nearly doubled since World War I, much of the emphasis on faster service has developed only since World War II. Since 1945 many railroads have introduced fast freights with schedules that live up to their names. The New York Central's "Pacemaker" and "Early Bird" fast freights had schedules of nearly passenger-train speed by the time Alfred E. Perlman (1902–83) became president of the road in 1954. The improved freight schedules were made easier as the nation's lines spent millions of dollars to install Centralized Traffic Control on a total of nearly 37,000 miles of road by 1965. Improvements in roadbed and heavier rail also helped. By 1957 nearly two-thirds of the nation's trackage weighed at least 100 pounds per yard. Improvements and new methods of track maintenance, such as the use of track-laying machines, powered spike hammers, ballast tampers and cleaners, and detector cars for finding hidden flaws in rails, had brought increases in both economy and operating efficiency. The savings in labor made possible by the new machinery were enormous, and by 1965 the rail workers maintaining way and structures numbered only 95,000, or less than a sixth of the total railroad employees.

A unique railroad effort to keep freight business was the introduction of "piggyback" service, the intercity transportation of truck trailers on specially equipped railway flatcars. Sometimes called "trailer-on-flatcar" service, the new approach not only created new railroad traffic but also proved that railroaders could serve truck shippers. *Niles' Weekly Register* in 1833 noted an early instance of such service on the Baltimore & Ohio, but the practice was certainly not common until well into the present century. Half a dozen railroads were offering some form of piggyback service in 1953. James M. Symes (1897–1976), after he became president of the Pennsylvania Railroad in 1954, was an early booster of the new service. In 1957 the first trailer-on-flatcar shipment was sent from New York City to San Francisco, and by 1959 fifty-seven railroads were offering the service in every one of the forty-eight continental states. The Chesapeake & Ohio was adding its piggyback equipment to passenger trains, the Milwaukee was moving bulk mail from Chicago to Wisconsin in "Flexi-Vans" on passenger trains, and more and more new automobiles were being delivered by the new service. Total piggyback carloadings in 1960 were well over 550,000, a figure 37 per cent above 1959 and 106 per cent above 1958. Although this was still well under 2 per cent of the total rail freight traffic, the phenomenal growth of the new service made it one of the brightest spots in the railroad picture.

As the railroads after World War I sought to maintain and expand their traffic by improving and upgrading their equipment, they were increasingly forced to finance the new equipment with a different type of security, the equipment obligation. In this transaction the trustee or legal owner of the new equipment retained legal title until the railroad, with a series of regular installments, had fully paid for the equipment. Since the equipment was readily movable and could easily be sold to other railroads in the event of default, the equipment obligation featured a high degree of financial safety. Many railroads, such as the Illinois Central, were obtaining new equipment in this way in the early twentieth century, and equipment trust obligations had increased nearly fourfold by 1950. Between 1945 and 1955 new equipment obligations amounted to at least 40 per cent of all new railroad capital expen-

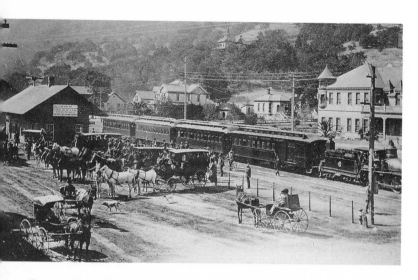

Passenger business was good in the 1880's when this Southern Pacific train stopped at Los Gatos, California. (Courtesy, Southern Pacific Lines.)

Van Buren Street Station, Chicago, Illinois, October 9, 1893, when a record 505,000 passengers, many of them visitors to the Columbian Exposition, rode the suburban trains of the Illinois Central. (Courtesy, Illinois Central Railroad.)

Railroad yard in 1917 during World War I. Congestion such as this was typical on eastern lines and hastened the operation of the railroads by the government. (Courtesy, Association of American Railroads.)

Two piggyback or trailer-on-flat-car trains meet on an Allegheny River bridge at Pittsburgh during their runs between New York and Chicago. The growing volume of this new freight service made it one of the brightest spots in the railroad picture in 1960. (Courtesy, Pennsylvania Railroad.)

ditures. Since World War II the great bulk of all new rolling stock and motive power has been financed in this way.

When the diesel locomotive appeared on the American scene, it not only brought a revolution to the nation's railways but also made possible tremendous increases in rail operating efficiency. The diesel engine was invented in the 1890's by French-born German mechanical engineer Rudolph Diesel (1858–1913). Adolphus Busch (1839–1913), a rich St. Louis brewer with modest interests in railroads and refrigeration equipment, bought the American manufacturing rights in 1898 but never applied the new engine to transportation. Some success with self-propelled rail cars driven by gasoline engines was achieved in the early twentieth century by several railroads, including the Union Pacific and the Santa Fe. When early diesel experiments failed, the General Electric Company solved the problem by combining diesel power with electric generators and traction motors.

In 1925 the first diesel-electric locomotive was installed in regular service by the Central Railroad of New Jersey for switching operations in New York City. Several dozen diesel switchers were in operation by the early thirties. In 1934, Ralph Budd, dynamic president of the Burlington, sought to retrieve some of the passenger business lost in the depression thirties by using diesel power for his new "Zephyr" streamliner passenger service. In the same year, Carl R. Gray (1867–1939), who was managing to make his Union Pacific pay dividends in spite of the depression, also used diesel locomotives for his streamliners. In 1941 the Santa Fe was the first road to use diesels for freight service, but before the end of the year it was joined by the Southern, the Great Northern, and the Milwaukee.

Diesel locomotives, or, more correctly, diesel-electric locomotives, offered a number of advantages over steam power. The new locomotives were admittedly expensive, costing, on the average,

$125,000 to $200,000 per unit. In cost per horsepower this was easily double that for a new steam locomotive. But the economic advantages of high fuel efficiency, low maintenance costs, and a high degree of availability offset these high original costs. At that time, the diesel hauled almost a ton of freight on a spoonful of oil, getting at least three times as much work from the same amount of fuel as the steam engine did. Since the diesel uses very little water, the nearly complete dieselization achieved by the late fifties permitted the retirement of $50,000,000 worth of water-supply equipment needed for the thirsty steamers. The maintenance costs of diesels proved very low. While the average steamer required extensive daily attention for its firebox and boiler, the diesel could run thousands of miles without having to be serviced en route.

Unlike the steam locomotive, which required an hour or two of "firing-up" before it could move under its own power, the diesel could turn on nearly full power from a cold engine almost instantly. The absence of reciprocating parts in the driving mechanism of the diesel also saved track from the pounding stresses typical of the steam engine. The fixed-horsepower capacity of the diesel did not lend itself to the overloading possible in the case of the more flexible steam engine, but an extra unit or two could be added to the diesel with little effort. Like electric locomotives on mountain grades, the diesel-electric could also reverse its electric motors to act as brakes.

By mid-century the diesel had so clearly proved its superiority that it was beginning to replace electric motive power as well as steam. Back in the 1890s experiments by the Baltimore & Ohio, and the New Haven with electrified train service were so successful that within half a century twenty railroads had electrified more than 6,000 miles of road. The Pennsylvania had over 2,000 miles of track thus operated, and the Milwaukee had more than 900 miles of the same type. Possessing all the advantages of the electric locomotive without the maintenance worries of electrified track, the diesel was clearly responsible for the decline from a high of 868 electric locomotive units in 1943 to less than 500 by 1960.

The greatest gains made by the diesel were at the expense of steam. The 1,200 diesel units of 1941 had grown during the war to 3,800 (out of a total locomotive roster of 43,500) in 1945, and in that year 7 per cent of the freight service, 10 per cent of the passenger service, and 25 per cent of the yard switching were performed by the new type of locomotive. In the decade after World War II the railroads of the nation spent $3,300,000,000 for 21,000 new engines, and 95 per cent of them were diesels. By 1955 more than fifty major railroads owned no steam locomotives at all. Not a single new steamer was bought by the nation's lines in the years 1954 through 1958. In 1957 the new form of motive power performed 92 per cent of the freight service, 93 per cent of the passenger service, and 96 per cent of the yard service.

In 1958 steam locomotives numbered under 1,300, many of them were in storage, and they handled less than 2 per cent of the freight service and under 1 per cent of the passenger and yard service. Even the Union Pacific's giant steamers, the 7,000-horsepower "Big Boys" of Sherman Hill fame, were in storage. At the end of the fifties the railroads possessed fewer than 30,000 locomotives (as contrasted to nearly 65,000 in the early twenties), but 96 per cent of them were diesels. By 1960 it was almost easier to find a steam locomotive in a museum than one operating on a railroad. The average child saw more of them in miniature under the Christmas tree than anywhere else. The obvious economies of dieselization had forced the raw, impersonal bleat of the diesel to replace the nostalgic whistle of the steam locomotive, a sound that had meant much to many Americans.

In the middle decades of the twentieth century other significant changes appeared in railroad equipment and rolling stock. In freight service the changes in cost (from $2,000 for an average boxcar in 1924 to $8,900 per boxcar thirty-five years later) were greater than those seen in the basic dimension, type, or style of the equipment. In the years after World War I, greater changes were made in passenger equipment as rail management sought, with a program of modernization, to meet the competition of bus, plane, and private automobile. Both the Pullman Company and Daniel Willard's Baltimore & Ohio were experimenting with mechanical air-conditioning for passenger equipment in the late twenties. Although the costs for such installments might average more than $5,000 per car, many roads introduced such service in the depression thirties, and by the fifties nearly all passenger cars were thus equipped.

Patents for streamlined trains had been issued just after the Civil War, and in the nineties inventor-author Frederick Upham Adams (1859–1921) was issued several such patents. Adams even persuaded the Baltimore & Ohio in 1900 to build in its company shops the Adams "Windsplitter," a train which in brief tests reached a speed of over 85 miles an hour. But serious streamlining waited for the depression thirties. When Ralph Budd and Carl Gray introduced diesel power in their passenger service in 1934, they also introduced streamlined passenger trains. The Burlington's stainless-steel "Zephyr" and the Union Pacific's "City of Salina," built of a lightweight aluminum alloy, both caught the fancy of a public which liked sleek and noiseless new trains which were free of smoke and cinders. In May 1934, the "Zephyr" set a world record with a 1,015-mile non-stop run between Chicago and Denver at an average speed of 77.6 miles per hour.

During the thirties, as many lines introduced streamliners, a new emphasis was placed on speed. In 1930, passenger trains operated at average speeds of 60 miles per hour for only 1,100 miles daily. A decade later, on the eve of World War II, passenger trains operated at those same speeds, or better, for 75,000 daily miles. As the public in the thirties expressed a preference for the new fast streamliners with smooth roof lines, shrouded wheels, and rounded rear ends, railroad after railroad introduced name trains of the new type. In May 1936, Samuel Thomas Bledsoe (1868–1939), president of the Santa Fe and another railroad man who believed in streamlined modern passenger service, helped inaugurate the fast "Super Chief" in its Chicago to Los Angeles run. The train, pulled by a new, gaily painted 3,600-horsepower diesel, introduced regular 39-hour service between the two cities, a schedule that for the first time broke the 45-hour record run of Death Valley Scotty back in 1905. In the forties the New York Central had nearly three dozen name trains on its main route, and just after World War II, in 1946, it announced a $56,000,000 order for 720 shiny new passenger cars. More than 350 name trains were in operation in the country by mid-century. But the new trains cost money. In 1952 the Pennsylvania Railroad pointed out that a new train for its New York City–to–Washington service, "The New Congressional" (electric locomotive and 18 coaches), had cost $3,224,000, or an investment of nearly $5,000 for each passenger. Back in 1914 an earlier "Congressional" (steam locomotive and 7 cars) had cost only $151,000.

The push toward modernizing and upgrading passenger service continued after World War II. One of the most insistent advocates of a new approach in railroading was Robert R. Young (1898–1958), a native of Texas weighing only 135 pounds but possessing a flair for headlines and a skill in management reminiscent of Commodore Vanderbilt. After a business apprenticeship with Du Pont and John J. Raskob (1879–1950), Bob Young made his first million during the dismal early thirties and by the early forties controlled the Chesapeake & Ohio system. In 1946 the Chesapeake & Ohio, which had very little passenger business itself, startled the railroad world with a series of newspaper and magazine advertisements which pointed out that a hog could travel across the nation without changing cars but that a human being could not. Chicago, where 500,000 Pullman passengers had to change trains in 1945, was an invisible barrier and a "phantom Chinese wall which split America in half." Young had made his point. Soon several railroads announced that they would book Pullman passengers through the Windy City without a change of cars. Another railroader with imagination, John W. Barriger (1899–1976), took over the Chicago, Indianapolis, and Louisville (renamed the "Monon" in 1956) in 1946. In six years he gave the short 600-mile Hoosier line a thorough face lifting.

Passenger equipment boasted new features, and sometimes a new silhouette, in postwar years. The idea of a "vista-dome" on a passenger car had been suggested in 1891 in the pages of *Scientific American*, and a dozen years later a dome of sorts was actually tried out on a passenger car on the Canadian Pacific. But real acceptance of the idea came only after the Burlington had tried such a car between Chicago and Minneapolis in July 1945. The public liked the new cars, especially along the scenic routes of the mountain West. Within a dozen years fourteen other railroads were using them, and more than two hundred such cars were in service. Fred G. Gurley (1889–1976), president of the Santa Fe and proud of the fact that his line was one of the three biggest users of diesel power, thought so much of the dome idea that in 1956, with his "El Capitan," he introduced the new concept of a "high-level" train. The new train provided an unobstructed view for all passengers as they rode at dome-car height. Baggage storage, larger rest rooms, and other services were on the lower level.

At the same time, the New York, New Haven & Hartford, James M. Symes's Pennsylvania, and the New York Central (where Robert Young had become chairman of the board in 1954) were all excited about several new lightweight low-level trains, such as Pullman Standard's "Train X," but the riding public soon expressed a preference for Mr. Gurley's approach to passenger-train travel. The public also liked the Burlington experiment in 1956 with "slumbercoaches," cars which provided sleeping accommodations for little more than the price of a coach ticket. The new service was so popular and so profitable that by 1959 the New York Central, the Northern Pacific, the Baltimore & Ohio, and the Missouri Pacific had all introduced comparable economy sleeping coaches.

In spite of organized research and expanded Centralized Traffic Control, increased efficiency and nearly total dieselization, streamlining and the appeal of piggyback service, railroad traffic continued to show a serious decline in the middle decades of the twentieth century. Despite the best efforts of railroad managers with imagination and the modern outlook, the railroad's share of commercial passenger and freight traffic continued to shrink. The convenience of the truck, the speed of the airline, the economy of the bus, and the cheap reliability of the pipeline contrived to reduce the railroad's share of the nation's transportation pie. The decline is easily seen in tables 8.5 and 8.6.

As the total American economy expanded in the twentieth cen-

TABLE 8.5

DISTRIBUTION OF INTERCITY FREIGHT TRAFFIC IN THE UNITED STATES

(BILLIONS OF FREIGHT TON-MILES AND PERCENTAGE OF TOTAL)

Year	Rail-roads	Per-cent-age	Inland Water-ways	Per-cent-age	Truck	Per-cent-age	Pipe-line	Per-cent-age	Air	Per-cent-age	Total
1916	367	77.2	88	18.4	...	...	21	4.4	...	...	476
1930	390	74.3	86	16.5	20	3.9	28	5.3	...	...	524
1940	379	61.3	118	19.1	62	10.0	59	9.6	.01	...	618
1945	690	67.3	143	13.9	67	6.5	127	12.3	.09	...	1,027
1950	597	56.2	163	15.4	173	16.3	129	12.1	.32	...	1,062
1955	631	49.4	217	17.0	226	17.7	203	15.9	.48	...	1,278
1957	627	46.4	232	17.2	260	19.2	233	17.2	.60	...	1,352
1960	579	44.1	220	16.8	285	21.7	229	17.4	1.00	...	1,314

Railroads in Decline

TABLE 8.6

DISTRIBUTION OF INTERCITY COMMERCIAL PASSENGER TRAFFIC IN THE UNITED STATES

(BILLIONS OF PASSENGER MILES AND PERCENTAGE OF TOTAL)

Year	Rail-roads	Per-cent-age	Inland Water-ways	Per-cent-age	Bus	Per-cent-age	Air	Per-cent-age	Total
1916	42	98.0	.5	2.0	...	...	...	...	43
1930	29	74.6	3.0	7.1	7	18.1	.07	.2	39
1940	25	67.1	1.0	3.6	10	26.5	1.0	2.8	37
1945	94	74.3	2.0	1.6	27	21.4	3.0	2.7	126
1950	33	46.3	1.0	1.7	26	37.7	10.0	14.3	70
1955	29	36.5	2.0	2.2	25	32.4	23.0	28.9	79
1957	26	31.9	2.0	2.4	25	30.6	29.0	35.1	82
1960	21	27.2	2.7	3.4	20	25.5	34.0	43.5	78

tury, intercity freight traffic nearly tripled in the four decades after World War I. Railroad freight in the same years far less than doubled. By the middle fifties the railroad's share of intercity freight traffic was less than half the total, the remainder being rather evenly divided among three major competitors. Railroad freight traffic declined at least 10 per cent since 1945. By contrast, that of inland waterways climbed 60 per cent, pipeline traffic nearly doubled, and the trucking industry increased nearly four-fold. In the sixties several aircraft companies had freight-carrying planes capable of carrying cargoes of from 65,000 to 100,000 pounds across the nation in from five to eight hours at an estimated cost of three to four cents a ton-mile.

The losses suffered in passenger traffic were even more severe than those in railroad freight. From a position of having a near monopoly of intercity passenger traffic before World War I, the railroads had declined by mid-century to only a third of the commercial traffic and less than 4 per cent of the total intercity traffic (including private-automobile mileage). Since the early days of railroading, freight revenue had been much more important than passenger revenue, but the passenger dollar became important in the early twentieth century. In the early twenties passenger revenue was roughly a quarter of freight revenue. By the early thirties it had dropped to a seventh and then during the gas-rationing days of World War II had climbed back to a quarter. In the fifties it averaged less than a tenth of the freight revenue, and by 1959 it had dropped to 8 per cent.

The decline in rail travel brought deficits. Since the early depression the nation's rail passenger service has resulted in annual deficits every year except for the four war years. These deficits climbed to half a billion dollars yearly by 1948 and reached a record high of $723,000,000 in 1957. Passenger service had cost the railroads of the nation, on the average, $1,400,000 a day—every day—since World War II. The annual passenger losses often came close to 50 per cent of the net railway operating revenue gained from freight service. While critics have sometimes looked askance at these figures, which were assembled from reports of the Interstate Commerce Commission, no one can deny that the losses from railroad passenger service were substantial.

Passenger deficits naturally brought a curtailment of service. In the twenties, when passenger service was still profitable, passenger trains operated, on the average, over nearly 225,000 miles of road, or on about 90 per cent of the nation's rail system. By 1939 passenger service was limited to only 171,000 route miles, and after World War II it declined even more rapidly. In 1965 passenger trains were running on 77,000 miles of line. No passenger service of any kind was thus available on more than half of the national rail network. Early in 1959 at least eleven Class I railroads had no passenger service at all. As short-line and branch-line service was abandoned, many passenger stations were closed or consolidated. In the state of New York in 1934 the New York Central had 235 passenger stations; in 1959 this figure had been more than cut in half, with only 113 still open.

Not only the extent but the density of passenger train service was drastically curtailed. This was especially true with eastern roads, where the decline in rail travel seemed to be more severe. One hundred long-distance trains pulled out each day from the Grand Central Terminal or the Pennsylvania Station in New York City in 1958. A decade earlier, 140 trains left daily for distant destinations. Between 1946 and 1958 the New York Central removed four of its six fast daily trains between New York and Chicago. The Pennsylvania cut its daily trains terminating in Washington from 26 to 14 in the same years. In 1957 the Pennsylvania gave up its Harrisburg–to–Washington service, saying, in effect, to the Baltimore & Ohio: "You can haul the passenger from Washington to Chicago." The B.&O. replied with a successful appeal to the regulatory authorities for permission to drop its service between Washington and New York. In the five years between 1951 and 1956, some 1,274 passenger trains were eliminated across the country. As the curtailment of train service made the typical railroad timetable grow thin, the traveling public became even more addicted to the joys of driving, the economy of the bus, or the speed of the airplane. Railroad management sometimes grew a bit irritated with a public which crowded its depots looking for stand-by rail service only when the airplanes were grounded or the highways were icy or fog bound.

Railroad managers also complained that they remained subject to the same intensive national and state regulation of services and fares as in the days when they indeed possessed a virtual travel monopoly. A railroad could not close a depot, discontinue a train (especially a "last train"), or hike a fare without regulatory approval. Most requests for rate increases were trimmed. The hearings and decisions were always slow, since, as a Boston and Maine executive put it, "Every rate case has become a carnival of oratory." Regulatory agencies were often unsympathetic to railroad requests for the discontinuance of trains which nobody really needed or used. For nearly thirty years the Milwaukee failed in its efforts to take off an Indiana train where the monthly revenues finally were less than a day's wages for the train crew. In New York State the Public Service Commission refused to permit a train to be dropped where the average trip carried only 4.5 passengers served by a crew of five. When the Transportation Act of 1958 was passed, Congress went a long way toward rescuing the passenger train from local politics. This Act gave the Interstate Commerce Commission both original and appellate jurisdiction over train discontinuances. The relief afforded was partly responsible for the reduced passenger-service deficit of $392,000,000 in 1962, the lowest in ten years.

A major reason for passenger-train deficits was simply rising costs, costs rising faster than revenue. Between 1948 and 1956 average expenses per passenger-train mile rose 39 per cent, while passenger fares rose only 15 per cent. The high cost of labor, or, from the viewpoint of management, the failure of railroad operating personnel to give a day's work for a day's pay, clearly contributed to the high costs. Following work rules unchanged since 1919, passenger-locomotive crews received a day's pay for every 100 miles run, and conductors and trainmen received the same for each 150 miles. On a sixteen-hour run between Chicago and Denver a name passenger train stopped about every two hours, the chief reason being to change train crews. The crew members called it a day as they received, on the average, more than a day's pay for about two hours' running time.

In 1961, the chairman of the Eastern Railroad Presidents Conference, David I. Mackie (1903–86), pointed out that road passenger-engine crews and freight enginemen (a total of about twenty thousand men) received more pay per hour worked (from $5.71 to $6.42 an hour) than did the nation's seven thousand top railroad executives ($5.57 an hour). In the years since World War II, many passenger trains were discontinued simply because the railroad brotherhoods rejected all proposals looking toward crew reduction. At that time, some of the operating personnel suggested that their unions would be willing to give in on the "featherbed-ding" issue if the railroads would pay overtime for night and Sunday work and give railroad labor a guaranteed annual wage.

The railroad passenger problem was also complicated and plagued by such thorny problems as the dining car, mail service, and the commuter. Even though the meals served are spartan compared to the nineteenth-century menu, railroads have not made any money on their dining-car services since World War I. Since Pullman patrons have deserted the trains even more rapidly than coach passengers in recent years, the deficits from the railroad dining car have grown accordingly. By 1954 it was costing the railroads $1.44 for every dollar they received for the high-priced victuals served to their patrons. In 1957 it cost $91,000,000 to operate the railroad's 1,400 diners. The year's revenue was only $62,000,000. Such innovations as pre-cooked meals offered on short runs, the American Plan meal coupons on the Atlantic Coast Line, or the ten-dollar meal ticket for five meals on the Santa Fe's "El Capitan" did little more than slow the rate of increase in deficits. Many roads, such as the Northern Pacific (the "Road of the Great Big Baked Potato") merely chalked the losses up to good public relations. They agreed with Harry A. de Butts (1895–1983), president of the Southern Railway, who insisted: "There is no way that you can lose or pick up freight customers so fast as by the quality of the meal or the cup of coffee that you provide the traveler."

The commuter and mail problems were just as serious. Even though the annual number of passengers commuting by rail in 1959 was only half that of 1929, they still constituted a grave financial drain on the twenty Class I railroads involved. Commuter traffic accounted for nearly a fifth of total passenger-miles and about a seventh of all passenger revenue. Though fares for city commuters have risen steeply since World War II, they still get a bargain compared to long-distance travelers. The heart of the commuting problem is the fact that the railroads are forced to maintain large rosters of expensive equipment for what is essentially a "twenty-hour week" consisting of a couple of two-hour peak periods five days a week. The outlying shopping center, television, and the automobile have combined to rob the railroad of most of its off-hour, evening, and weekend suburban traffic. Substantially, to hike the fare seems unfair to the commuter, since he is being asked to pay the full cost of a hopeless and endless contest between a tax-paying corporation and a tax-free, community-supported highway facility. As city planners come to realize the economies of speed and space available in railroad commuter service as compared to that of the private automobile, more communities are apt to give their railroads the same relief in local taxes as that granted to the Long Island Railroad. The railroads are also bothered by an experiment started by the Post Office Department in the middle fifties in which first-class mail was sent by air whenever there was available space in mail-carrying planes. The railroads fear that as passenger service is further reduced, they may ultimately be left with only the bulky and less profitable second-, third-, and fourth-class mail sacks.

Thus American railroads must look back upon many decades of relative decline, decades in which their best efforts seemed only to slow the success of the new competitive transport facilities. The battle of traffic statistics has consistently been lost to the private auto, the bus, the truck, the plane, and the pipeline, even though the railroads have fought back with diesels, piggyback service, Centralized Traffic Control, slumber coaches, and faster schedules. Without the notable increases in operating efficiency gained in the years since World War I, the competitive task of the railroads might well have seemed hopeless. Just since World War II, the railroads of the nation put over $14,000,000,000 in new capital improvements, and yet from 1945–60, they averaged a return of little more than 3.5 per cent on their net investment, a figure well below that of all other major industries. In 1958 and 1959 the railroad rate of return was no higher than 2.8 per cent, and in the latter year, nineteen major lines actually operated in the red. Neither financial circles nor the general public could believe that American railroads had adjusted themselves too well to the transportation revolution of the early twentieth century. There was too much contrast between the airlines' sleek and shiny new jets and the railroad passenger coaches which in 1960 had an average age of twenty-nine years.

The poor health of the railroad industry could hardly be denied when one noted that the 1959 rail mileage was less than that of 1906, although capitalized at more than twice as much; that railroad passenger traffic in 1959 was below the level of 1905 and yet received an average fare that was only a penny per mile higher; that railroad employment in 1959 was below the level of 1896, with the average annual wage up nearly elevenfold. The golden age that American railroads had known in the generation before the First World War had, by the sixties, turned a sickly yellow.

9

Troubles in the 1960s and 1970s

American railroads were facing a variety of problems in the decade of the 1960s. A low rate of return, a real or relative decline in freight traffic, passenger service deficits, irksome federal regulations, "featherbedding" by the work force, and a poor image held by the public were all vexing to railroad management. During the 1950s the average rate of return for railroad property was only 3.67 per cent. Between 1960 and 1964 this figure dropped to 2.62 per cent, and railway shareholders received only 2.23 per cent. The rate of return showed no real improvement in the late 1960s or the decade of the 1970s. Throughout the twenty-five years, northeastern roads were less prosperous than those in the South and the West. The volume of rail freight dropped sharply in the late 1950s and did not recover for a decade. The recession of 1957–58 was partly at fault, but the major cause was the expanding truck traffic on the newly opened Interstate Highway System. The share of intercity freight moving by railroad had dropped from 49.5 per cent in 1955, to 44.1 per cent in 1960, to 39.6 per cent in 1970, and to 37.5 per cent in 1980.

In the same decades the share of intercity freight moving by highway trucks had grown from 17.5 per cent in 1955 to 21.3 per cent in 1970 and to 22.3 per cent in 1980. Rail freight had also been lost to rivers, canals, and pipelines. As the volume of railroad freight slowed, the mix of shipments on steel rails also changed. More and more farm produce, household goods, and manufactured items began to move on rubber and concrete. A great advantage for the highway truck was the convenience of door to door service. In 1967, Thomas M. Goodfellow, the president of the Association of American Railroads, said that his industry had to make itself: "lean, hard, and efficient" to meet the new challenges.

The slump in railroad freight varied among the regions of the nation. The Interstate Commerce Commission had divided the country into three districts: Eastern, Southern, and Western. The Eastern District was the trunk-line area—east of St. Louis and Chicago, and north of the Ohio River and North Carolina. The Southern was east of the Mississippi and south of the Ohio River and Virginia. The Western was west of Lake Michigan and the Mississippi. Freight ton-mileage for the three districts are shown in table 9.1.

TABLE 9.1

RAILROAD FREIGHT TON-MILES (IN BILLIONS)

Year	United States	Eastern District	Southern District	Western District
1955	624	257	90	277
1960	572	218	88	266
1965	698	259	117	322
1970	765	254	140	371
1975	754	218	140	396
1980	919	202	170	547

Between 1955 and 1980, the total rail traffic in the United States had modestly increased, that in the Eastern had slightly declined, the Southern had climbed 88 per cent, and the Western had

nearly doubled. Eastern railroads had faced several unique problems in these years. Dozens of eastern factories and plants had
moved their operations to the south or to the west to escape the
high wages of unionized labor. Northeastern lines were also handi-
capped by expensive congested terminals, shorter average freight
hauls, and the high cost of providing commuter service in metropolitan areas. Most eastern railroads had higher than average operating ratios. Many eastern lines, including James M. Symes's
Pennsylvania, continued to pay regular dividends in spite of declining freight and passenger traffic.

The continuing decline in passenger traffic was of concern to
all railroads in the 1960s. The Transportation Act of 1958 had permitted faster ICC action in the discontinuance of passenger service, and in the next few years nearly a thousand passenger trains
were dropped. Alan S. Boyd, Secretary of Transportation, in referring to the decline of rail passenger service, said: "Andrew Jackson
was the first President to ride a train. We think one of our respon-
sibilities at the Department of Transportation is to see that President Johnson is not the last." In 1958, passenger trains were in
service on 107,000 miles of line. By 1970, such service existed on
only 49,000 miles of the rail network of 206,000 miles. Many railroads had no passenger trains at all in the late 1960s. In 1966 Indianapolis had 18 daily passenger trains, less than a tenth of the 184
trains serving the city fifty years earlier. A few states, such as
Maine, feared the loss of all passenger service. Passenger-miles of
travel on American railroads declined from 21.6 billion in 1960 to
10.9 billion in 1970.

In explaining the drop of passenger service on the Santa Fe,
president of the road, John S. Reed, in 1967, said: "Santa Fe has
not abandoned the traveling public—travelers show an increasing
preference to drive or fly." High labor costs on passenger trains
were a major problem in the sixties. William J. Quinn, president
of the Burlington in the late 1960s, pointed out that a Boeing 727

in its two-hour Chicago to Denver flight had labor costs of $391 for its six-member crew. The California Zephyr in its 18½-hour Chicago-to-Denver run had a wage bill of $2,288 for its forty-seven crew members, who had frequent crew changes. The airlines made a profit of $943 for its trip, while the Zephyr had a loss of $334. Railroad annual passenger service deficits were indeed high, with an average loss of nearly half a billion dollars from 1958 to 1970. During these years the passenger deficit cut the industry's net income by a third or more. By 1970 the volume of railroad passenger traffic was less than half that of the intercity bus, and less than a tenth that of airlines. A different plan was needed to provide rail passenger service for the nation.

Railroads in the 1960s were also unhappy with the rigorous federal regulation, their image with the public, and the high taxes they paid. President Dwight D. Eisenhower (1890–1969), late in his term in the White House, said of the railroads: "I think they are governed by antiquated laws and regulations, and frankly, I think some of their trouble is their own." Certainly the hand of regulation was heavy. The ICC required the railroads to keep 258 types of records, while airlines had to keep only a fifth as many. Railways had to maintain statistics on the cost of sledges, scythes, and sickles, prorating the cost between freight and passenger service.

Railroads rated poorly with the public. In the 1960s the American public, completely enchanted with the jet-age of air travel, viewed railroads as old-fashioned and the original pioneers of pork barreling back in the days of huge land grants. Thomas Goodfellow urged his colleagues to: "Shout loudly . . . and weep no more." But railroads could afford few dollars for public relations. In 1968 the entire industry had only 20 pages of advertising in the *New York Times*, while the airlines had 791 pages. Between 1967 and 1981, taxes paid by the nation's railroads more than tripled. The railroads, of course, provide their own depots and terminals, which are taxable. In 1961 the tax paid on the Union Station in Washington, D.C., was about equal to the deficit the federal government picked up that year for Washington National Airport. David I. Mackie, chairman of the Eastern Railroad Presidents Conference, noting that highways, canals, and airways were free of taxation for their route, suggested that the states should eliminate taxes on railway rights-of-way. However, few state tax officials agreed. The public reaction to any suggestion for any aid for the railroads was apt to be, in the words of James M. Symes of the Pennsylvania, "as if burglars had asked for changes in laws against safecracking or housebreaking." When the New York Central was forced to build a $23,000,000 bridge over the Harlem River to provide clearance for a new highway underneath, New York raised taxes on the improvement from $70,000 to $500,000.

In the 1960s, railroad managers did face up to the problem of "featherbedding" by railroad labor, especially the 188,000 engineers, firemen, conductors, and brakemen who ran the trains. These members of the Big Four Brotherhoods had average annual wages of about $8,000 in 1961. Such a figure was probably only half the average physician's salary, but would easily match that of a college professor that year. Railroad officials at a Chicago labor meeting told each other about the passenger train engineer out in Kansas who had so lush a run that eighty hours in the engine cab earned him $800 a month, giving him plenty of time to manage a 700-acre farm, which was yielding $800 a month from the government for not growing wheat. Daniel P. Loomis, head of the Association of American Railroads, claimed that "featherbedding" cost his railroads $500,000,000 a year, while Henry E. Gilbert, head of the Brotherhood of Firemen, contended that "featherbedding" was common in all American industry.

As early as 1959, Wayne A. Johnston (1897–1967), president of the Illinois Central, was calling the diesel fireman's shovel an "expensive antique" since he had no coal to shovel. Earlier a Canadian Royal Commission had decided that diesel firemen were not needed, and some European nations had removed diesel firemen from passenger trains. The Association of American Railroads insisted on the change. Between 1959 and 1963, several federal commissions arbitrated the dispute. All the commissions decided in favor of the railroad managers. The U.S. Supreme Court in a 1964 decision held that all but 10 per cent of the diesel freight and yard firemen should be eliminated. Rail management did not have comparable success in lengthening the "work day" of 100 miles for freight train crews and 150 miles for passenger train crews. It was fortunate for railroad management that when the Big Four Brotherhoods insisted on having firemen on the first diesels, they had permitted one or more extra diesels (in a long freight train) to be operated by a single engineer in the leading diesel unit. The total engine and train crew roster declined from 188,000 in 1961 to 164,000 in 1970, and to 136,000 in 1980.

Many rail officials in the 1960s, and even before, believed that merging with other lines might bring a cut in mileage, a reduction in work force, an increase in general efficiency, and a lower operating ratio. In 1957 about 50 merger applications had been proposed to the Interstate Commerce Commission. The ICC approved less than half of the requests.

In 1955 the Louisville & Nashville asked the ICC to approve its merger with the 1,043-mile Nashville, Chattanooga & St. Louis. Approval was given in 1957. Two years later, in 1959, the Norfolk & Western, one of the major eastern coal roads, gained approval of the ICC to take over the Virginia railway, a 600-mile line with very low grades over which millions of tons of coal moved from West Virginia mines down to Norfolk each year. The acquisition had been made by Stuart T. Saunders (1909–1987). Saunders was a lawyer and represented a new breed of railroad executives who moved up through legal and financial channels, rather than that of more traditional railway operations. In 1960 the 960-mile Delaware, Lackawanna & Western (Lackawanna) and the 2,200-mile Erie completed a merger, taking the name Erie-Lackawanna Railroad. Both original lines served common territory in New Jersey, Pennsylvania, and New York, with the Erie system running west to Cincinnati and Chicago. The new Erie-Lackawanna managed to abandon some duplicate mileage, but the railroad did not prosper in the 1960s.

A much larger merger took place in the early 1960s, when the 5,200-mile Chesapeake & Ohio acquired the 5,900-mile Baltimore & Ohio. In the 1950s the B&O had suffered a drop in revenue and was facing a troubling deficit. In contrast, the C&O, under its skillful president, Walter J. Tuohy (1901–1966), was easily paying $4.00 dividends. Tuohy, who had been a major coal company executive, was long on drive and imagination. In the last weeks of 1960, Alfred E. Perlman, president of the New York Central, was also eagerly buying B&O stock, but by February, 1961, Tuohy held a majority of the B&O shares.

The ICC approved the C&O acquisition of the B&O in early February 1963. The merger was successful, partly because the two roads tended to be "end to end" rather than parallel. After Tuohy's death, Hays T. Watkins, Jr., vice-president of finance, soon became the dominant figure in the Chesapeake System, which was formed in 1973. The C&O-B&O had added the 808-mile Western Maryland in the late 1960s. Further south the 5,200-mile Atlantic Coast Line and the 4,000-mile Seaboard Air Line, two major southeastern roads, started merger talks, which were approved with the creation of the Seaboard Coast Line in mid-1967. This powerful new competition in the south had caused the 7,600-mile Southern Railway to acquire the 1,800-mile Central of Georgia in 1963. A year later, in 1964, the ICC approved the earlier request by Stuart Saunders that the Norfolk & Western acquire both the Wabash and the Nickel Plate, two moderate sized roads serving mid-America from Buffalo westward to Chicago, St. Louis, and

Kansas City. The ICC finally approved this merger, but it insisted that the Norfolk & Western take over two weak lines, the Delaware & Hudson and the Erie-Lackawanna.

The largest of the railway mergers in the 1960s was that of the New York Central and the Pennsylvania. In the mid-1950s these two lines were arch rivals as well as the largest American railroads. Their combined mileage of 21,000 miles was only 9 per cent of the national total, but their main lines and dozens of branch routes produced about a fifth of the total rail revenue in the nation. The first talk of such a merger was in 1957 between Robert Young, chairman of the Central, and James Symes of the Pennsylvania. News of such a merger surprised the industry, since it had long been thought that the northeastern lines would some day arrange themselves into two giant systems headed by the two rivals. The formal merger request to the ICC was made in 1962. When Stuart Saunders became president of the Pennsylvania in 1963, he pushed for consolidation with vigor, convincing Alfred Perlman, by then the Central president, that the merger would eventually produce yearly savings of $80,000,000. However, the proposed union was to be a joining of contrasting lines. The Central claimed a water level route—the Pennsylvania followed a shortcut through the mountains. The Pennsylvania favored tuscan red and 4-6-2 Pacifics, and the Central preferred gray and 4-6-4 Hudson steamers.

The ICC finally approved the merger and the creation of the Penn Central, which was confirmed in an opinion of the U.S. Supreme Court in February 1968. However, the ICC had stipulated several conditions: (1) the Penn Central would take over the bankrupt New Haven line, (2) it would either retain or "buy off" many redundant current employees, and (3) it would give up its stock control of the Norfolk & Western. Saunders became chairman, and Perlman president, of the new Penn Central. Very little planning had been done by either line before the approval of the merger. Saunders took much of the money from the sale of the Norfolk & Western stock and invested it in rather questionable conglomerate business ventures.

The new giant Penn Central system faced a variety of problems. One of Saunders' aides, Allen J. Greenough (1905–1974), said of the enlarged line: "This is a big dog with a lot of fleas . . . We'll be scratching for a long time." The Penn Central was officially one railroad, but Central workers tried to follow their old rules, and Pennsylvania workers did likewise. The two old lines had very different signal and computer systems. The result in 1968 and 1969 was lost waybills, lost freight shipments, and even one lost train. By early 1970 a major problem was having to borrow capital at 8 to 10 per cent for a railroad earning 1 or 2 per cent. Penn Central was now losing one million dollars a day. On Sunday, June 21, 1970, Penn Central filed for bankruptcy, one of the biggest business failures in United States history.

Between 1970 and 1976, Congress faced up to two major railroad problems: the decline and collapse of rail passenger service, and the bankruptcy of the Penn Central. The first action came in October 1970, when President Richard M. Nixon (1913–1994) signed the legislation which set up the National Railroad Passenger Corporation. The Act was to relieve the railroad of an ailing and deficit-ridden passenger service. The general public was concerned about the expanding list of train discontinuances—from 20,000 intercity trains operating in 1929 down to only 500 in the fall of 1970.

On May 1, 1971, the National Railroad Passenger Corporation opened a 21,000-mile passenger service network called Amtrak, a contraction of two words—"America" and "track." The network extended from Seattle to Miami, and from San Diego to Boston. But Amtrak served only 43 states (there was no service in Maine, New Hampshire, Vermont, Arkansas, or South Dakota). Over 400 different cities were to be served by Amtrak, but neither Dallas

(844,000 population) nor Cleveland (751,000) were included in the 1971 schedules. Both cities would later be provided with Amtrak service. The July 1971 nationwide schedule book consisted of 32 pages, a third fewer than the Pennsylvania Railroad schedule of October 30, 1960. The 21,000-mile Amtrak network was less than half the 1970 figure of 49,000 miles of passenger service. Amtrak was to be financed by passenger ticket sales, federal grants and loans, and onetime charges upon the 13 railroads being relieved of their unprofitable passenger service. The government hoped that ticket sales might cover one-half to two-thirds of the annual Amtrak expense. In 1971 the Illinois Central paid Amtrak $9,000,000, and the B&O paid $5,000,000 as onetime charges. The total charges paid by all the 13 lines came to more than $190,000,000. Three roads avoided paying by keeping some passenger service: Southern Railway (New Orleans to Washington), the Denver & Rio Grande (Denver to Ogden), and the Rock Island (Chicago to Rock Island). Amtrak ran its trains over the rails of 15 to 20 railroads, paying in long-term contracts for maintenance and other services.

Roger Lewis (1912–1987), former aviation and airline official, was president of Amtrak from 1971 to 1975. The first equipment for Amtrak—1,200 cars and 300 locomotives, a diverse mix of coaches, sleepers, diners, lounges, diesel units—was supplied by member railroads. The equipment had an average age of twenty years or more and was called the "Rainbow Fleet." The cars varied in design, electric voltage, and type of heat and air-conditioning. Mechanics were in despair when called upon to repair such a variety of equipment. Amtrak officials made no effort to match the service or speed of the passenger schedules of the earlier 1950s. The public was soon complaining that Amtrak trains were rarely on time. In 1972 Amtrak had an operating revenue of $163,000,000 and expenses of $310,000,000 in providing three billion passenger-miles of service.

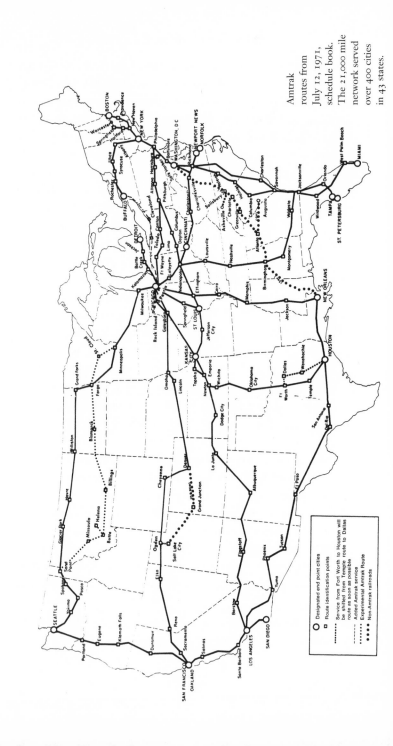

Amtrak routes from July 12, 1971, schedule book. The 21,000 mile network served over 400 cities in 43 states.

In 1975, Paul Reistrup, a railroader and former Illinois Central vice-president, replaced Roger Lewis as president of Amtrak. The idea of high-speed passenger service in the Northeast had originated earlier with congressional passage of the High-Speed Ground Transportation Act of 1965. In 1976, under Reistrup's leadership, Amtrak acquired from the defunct Penn Central and New Haven the 456-mile multi-track line from Washington, D.C., via Baltimore and New York City to Boston. Amtrak soon developed a faster Metroliner service over this Northeastern Corridor route. Faster service on the Corridor route was possible because Amtrak trains were free of much of the former freight traffic. Much new equipment was acquired in the mid-1970s for the entire Amtrak system. Several hundred F-40 diesel units were acquired from General Motors' Electro-Motive Division. Also, many double-decker Superliners were obtained from Pullman Standard. By 1980 the average age of Amtrak cars had declined to fourteen years and of locomotives to seven years. In 1978, Alan S. Boyd, former president of the Illinois Central, succeeded Reistrup as head of Amtrak. In 1979 Amtrak had revenues of $496,000,000 and expenses of $1,113,000,000 for 4.5 billion passenger-miles of service.

The other major railroad problem faced by Congress in the 1970s was the sudden failure of Penn Central in June 1970. As rail traffic in the late 1960s had declined in the Northeast, maintenance on the Penn Central had been cut back. This caused a deterioration in service which the bankruptcy finally disclosed. The failure of the Penn Central quickly affected connecting roads. The Lehigh Valley also failed in 1970, the Reading in 1971, and both the Erie-Lackawanna and the Lehigh & Hudson River in 1972. Bankruptcy and receivership does not mean that railroads quit operating—they simply are freed of paying their debt and fixed charges. Court-appointed trustees could find no buyers for segments of the Penn Central. The Penn Central service was so bad that some main lines in Ohio and Indiana had "slow orders" reducing speed to 10 miles per hour. Penn Central requested that all passenger service west of Buffalo and Harrisburg be discontinued, a ploy which may have hastened Congress to create Amtrak. With Penn Central losing $237 million in 1970, and $180 million in 1971, major creditors of the railroad demanded liquidation in the absence of any plan to reorganize the line.

There was a general agreement in 1972–73 that the economic health of the Northeast and Midwest required some reorganization of the Penn Central and its connecting railroads. Estimates appeared that a shutdown of Penn Central would affect nearly half of the nation's factories, throw hundreds of thousands out of work, and reduce the Gross National Product by 2 per cent within a few weeks. Frank Barnett, chairman of the Union Pacific, was the author of a plan to have the federal government reorganize the Penn Central. On January 2, 1974, President Nixon signed the Regional Rail Reorganization Act of 1973. This legislation called for the planning of a new regional railroad freight system to be set up as a profit-making system. The new Northeastern rail system would use $2 billion of federal funds to modernize right-of-way, track, and equipment.

Early in 1976, President Gerald Ford approved the Railroad Revitalization and Regulatory Reform Act of 1976, which in turn created the Consolidated Rail Corporation (Conrail), a private for-profit railroad whose capital stock was to be held by the creditors of the several bankrupt railroads to be aided. When Conrail began operation on April 1, 1976, it had 96,000 employees, 4,600 locomotives, and 152,000 freight cars. Conrail included not only Penn Central mileage, but more than 5,000 miles of line of five other nearly bankrupt roads. It served southern New England, the Mid-Atlantic states, and the Old Northwest except for Wisconsin. Conrail also served commuters in New York City, Philadelphia, Baltimore, Washington, D.C., and Chicago but was to be reimbursed for any losses by the cities it served. By 1978 Conrail had spent $2.1 billion of federal money and $490 million of private financing for rebuilding, but was still running in the red. In 1978 an average of 61 trains a day failed to move because no locomotives were available, and some 9,000 miles of track were under reduced special orders. Conrail would not become a profitable system until the early 1980s.

Conrail was to be the final solution to the failed Penn Central merger. But the decade of the 1970s would see several mergers both in the West and the South. The merger of the three "Hill roads"—the Burlington, the Great Northern, and the Northern Pacific—was finally approved by a 7–0 vote of the Supreme Court on February 2, 1970, four months before the failure of the Penn Central. The three roads were very different: the Burlington (CB&Q) was the strongest of the several "Granger" roads; the Northern Pacific had been given the most generous of the land grants; and Jim Hill had built the solid Great Northern without any help from Uncle Sam. Later, *Railway Age* would call the merger of the three roads into the Burlington Northern the "most positive development" of 1970.

Unlike the Penn Central, the Burlington Northern merger was preceded by years of careful planning. Plans had started as early as 1956, and the formal request to the ICC was made in 1961. In the 1960s John M. Budd (1907–79), president of the Burlington and son of Ralph Budd, and Louis M. Menk, president of the Northern Pacific, had worked diligently for the merger. With the addition of the Spokane, Portland & Seattle (long controlled by the Great Northern and Northern Pacific), the Burlington Northern, with 25,000 miles of line, was the longest railroad in the country. The success of the BN merger quickly brought distress to the Chicago, Milwaukee, St Paul & Pacific. The Milwaukee line could not meet the competition of the BN. The Milwaukee's final bankruptcy came in 1977: some 7,000 miles were abandoned, and some of the remaining routes were purchased by the Soo Line in 1986.

Other lines serving Chicago also were merging in the 1970s. In 1972 the 6,500-mile Illinois Central merged with the 2,700-mile Gulf, Mobile & Ohio to become the Illinois Central Gulf. Earlier, in 1963, the Illinois Central, under Wayne Johnston's leadership, had started to diversify, creating Illinois Central Industries, adding to the rail line a brake-shoe company, a soft-drink bottling works, and Chicago real estate. Several other railroads also tried diversification in the 1970s. Another Granger road, the Chicago & North Western, had a chairman, Ben W. Heineman, who not only followed the IC into diversification but allowed his workers in 1972 to become co-owners of the railroad. Back in 1961 the Union Pacific had proposed a merger with the Chicago, Rock Island & Pacific, a Granger line in some financial trouble. The ICC took thirteen years of hearings and delays to make up its mind, a fact which pushed the Rock Island into its third and final bankruptcy in 1975. The Rock Island shut down in 1980 and was liquidated. Further south the Missouri Pacific acquired the Texas & Pacific in 1976.

During the 1960s and 1970s there continued to be considerable downsizing in American railroads, both in mileage and employment. The 217,000-mile network of 1960 was 15 per cent below the top mileage of 1916. By 1970 the total was 206,000 miles, which dropped again to 179,000 miles in 1980. But the decline in employment was even greater. Railroad employment for Class I lines in 1960 was 780,000, down sharply from the all-time high of about two million in 1920. The total dropped to 566,000 in 1970 and 459,000 in 1980. Back in 1920 the railroads employed about 8 workers per mile of line—by 1960 the figure was a little over 3 per mile and just over 2 per mile in 1980.

In the two decades retail prices and the cost of living greatly increased. During the 1960s prices rose 2 or 3 per cent a year, but in the 1970s they climbed from 3 to 14 per cent annually. The higher inflation in the 1970s was largely the result of the oil and energy crisis and the new economic power of OPEC (Organization of Petroleum Exporting Countries). Between 1960 and 1980 consumer prices nearly tripled, climbing from 89 in 1960, to 100 in 1967, to 247 in 1980. In the same twenty years average yearly wages for Class I railroad workers increased from $6,270 in 1960 to $24,659 in 1980, almost a fourfold increase and well above the inflation rate. The average non-railroad worker in the 1970s did not receive wage increases that even matched inflation.

In 1980 the 17,000 rail executives (and staff) had an average yearly pay of $35,673, while the 136,000 train and engine crew members had an average pay of $29,101. Some through-freight engineers were making $38,000 to $40,000, while average pay for section hands was $18,000. In 1977 the labor columnist for *Railway Age* wrote rather wryly: "Railroad employees do not come off badly at all." By the late 1970s rail management had not yet had any luck in changing the sixty-year-old rule of a "day's work" for train crews: 100 miles for freight crews and 150 miles for passenger train crews. In 1976 railroad wages (not including fringes) came to 46.4 per cent of the revenue dollar. In the troubled Eastern District the figure was 56 per cent.

Inflation also hit railroad fuel and the cost of equipment. Diesel fuel for locomotives rose from 9 cents a gallon in 1960, to 11 cents in 1970, to 30 cents in 1975, and to 81 cents in 1980. The cost of new freight cars climbed. Between 1964 and 1980 the cost of a general service boxcar climbed from $13,000 to $46,000, flatcars from $17,000 to $50,000, and refrigerator cars from $22,000 to $70,000. Fortunately, average freight capacity increased from 55 to 79 tons between 1960 and 1980. The number of freight cars in service dropped from 1,965,000 in 1960 to 1,791,000 in 1980. One of the reasons for the increased cost of freight cars was the greater use of roller bearings. Passenger cars had used roller bearings in their car trucks for some time, but after 1963 all new freight cars had to have roller bearings instead of the old babitt-faced journal bearings. Roller bearings greatly reduced the number of "hot-boxes" that plagued bearings of the older type. By the mid-1970s about half of all freight cars had roller bearings. The index cost of railroad materials and supplies (other than fuel and freight cars) rose from 94 in 1960, to 100 in 1967, to 109 in 1970, and to 259 in 1980.

Even with the increased costs of labor, fuel, freight cars, and supplies, American railroads managed to restrain the rise of freight rates in the 1960s and 1970s. The freight rate per ton-mile was 1.40 cents in 1960, climbing only to 1.43 cents in 1970, and rising to 2.88 cents in 1980. Freight rates had only doubled in twenty years while fuel had climbed ninefold, the cost of labor had increased nearly fourfold, and freight equipment had roughly tripled in cost. The decline in work forces in the two decades, the acquisition of larger locomotives, the use of larger freight cars moving more miles per day (45 miles in 1960, 60 miles in 1980), and the abandonment of many miles of less-used track, all helped account for the restrained increase in freight rates.

Several other innovations in rail service either appeared or were expanded in the 1960s and 1970s. Unit trains are an example. Earlier "block trains" had moved bananas north at express train speed on the Illinois Central to the Chicago market in the 1880s, and "silk trains" had run eastward across the nation in special fast, well-protected shipments. The ICC had generally frowned on trainload rates, but the Commission in 1958 finally permitted lower rates for multiple-car shipments, as projected coal slurry lines and cheap oil threatened the railroad coal traffic. The new unit coal train avoided classification, weighing, and other terminal delays, and greatly improved car utilization. Dennis W. Brosnan (1903–1985), Southern Railway president in the 1960s, found that 200 unit train cars did the work of 700 earlier used hopper cars.

Some railroaders said a unit train was "one car repeated many times." The Denver & Rio Grande Western was an early user of unit coal trains. A division superintendent on the D&RGW said:

"All segments of the operation must interlock so that the total system functions like clockwork." Unit train service was further improved when the customers set up special high-speed loading and unloading equipment. With the increased availability of western coal from Wyoming and Montana, unit trains became even more popular. Both the Union Pacific and the Burlington Northern soon had dozens of 100-car unit trains moving coal 1,000 to 1,200 miles to eastern markets. By 1977, nearly half of the coal moved by rail was being carried on unit trains of eighteen different railroads. Conrail, Louisville & Nashville, and Norfolk & Western were other big users of this new type of service. Coal was the major reason for unit trains, but other commodities carried included grain, iron ore, and chemicals.

Piggyback service also grew in the 1960s and 1970s. Between 1960 and 1980 the cars of such service increased threefold with the greatest growth being in the Southern and Western districts. In the same years Alfred Perlman introduced the Flexi-Van on the New York Central, which eliminated the trailer chassis and wheels. Using a van or container, rather than a trailer, permitted a lower center of gravity and higher speed, as compared to regular piggyback service. In the same years when the Southern Railway also started to use containers, Dennis Brosnan had his engineering staff design and build huge cranes to load and unload the containers. The new types of service were known as TOFC (trailers on flatcars) and COFC (containers on flatcars). Container traffic grew rapidly, climbing from 590,000 revenue cars in 1965 to 1,350,000 cars in 1980. By 1980 the combined TOFC/COFC traffic was second only to coal traffic in total rail freight volume.

Unit trains, and to some extent TOFC/COFC traffic, were aided by track and rail improvement in these decades. Much of the new rail laid down in the 1970s was continuous welded rail, put down in lengths of 1,500 feet or more. The welded rail eliminated the familiar "clickety-clack," and produced estimated savings of

$1,000 a mile per year in maintenance expenses. By 1980, three-quarters of all rail on Class I lines weighed 100 pounds per yard or more. More and more improved track maintenance equipment appeared in the 1960s and 1970s. Not many years ago an experienced section hand could replace one tie in one hour. Today mechanized equipment can replace one tie in one minute.

Railroads were using more computers in these decades. By 1970 the Union Pacific was using 53 IBM computers in 38 yard offices to continuously monitor all their line operations. Computers were widely used in the accounting departments of all the railroads. Both computers and television were used in larger and more sophisticated Central Traffic Control offices. During World War II and after, Central Traffic Control was used for hundreds of miles of line. By 1980 it was used to control traffic for thousands of miles of line stretching over several states.

American railroads did not solve all the problems facing them in the 1960s and 1970s. Their share of intercity freight traffic continued to decline, but by the late 1970s seemed stabilized at about 37 per cent of the total intercity traffic. However, rail freight service grew more efficient in the twenty years. Thanks to the downsizing of the labor force and mileage, better and larger cars and diesels, and new types of service, the railroads greatly increased service per employee. In 1960 the annual freight service per worker was 733,000 ton-miles. Twenty years later, the figures had nearly tripled, being 2,100,000 ton-miles per employee. The 1980s and early 1990s would be even better years for American railroads.

10

The Staggers Rail Act of 1980 Brings Prosperity and Many Mergers

Toward the end of his presidency, Jimmy Carter favored legislation that would partially deregulate the railroads of the nation. On October 14, 1980, Carter signed the Staggers Rail Act of 1980. The legislation was named for retiring Harley O. Staggers (1907–1990), a Democratic Congressman from West Virginia, who was chairman of the House Commerce Committee, and famous for his insistence that Amtrak should adequately serve his home state. The 1980 Act was worked on by several other Congressmen, but Staggers was the official sponsor of the legislation. The supporters of the bill argued that railroads in 1980 were carrying only a large third of intercity freight, and that rail lines no longer had the transport monopoly that the ICC had been created to regulate. Railroads in 1980 had lower earnings than any other form of transportation, and were estimated to have inadequate assets to meet their needs for new capital in the early 1980s.

The Staggers Act loosened ICC regulations on rate making, marketing, mergers, and abandonments. The ICC was required to decide on applications for abandonments within nine months, where previously the Commission had no limits on the length of its discussions.

Railroads were given freedom to bid for traffic, or turn it away, on the basis of their own costs and to raise rates that fell well below their out-of-pocket costs. The Staggers Act also shifted the burden of proof in freight rate cases onto the shippers and away from the railroads. William H. Dempsey, president of the Association of American Railroads, wrote of the Act: "Deregulation lays the foundation for a new era of railroad growth and prosperity by granting railroads greater opportunity to respond to . . . the marketplace." Harold H. Hall, president of the Southern Railway, said: "No longer will we have the ICC to blame for our troubles. We will be masters of our own fate."

The 1980 legislation was followed a year later by the Northeast Rail Services Act of 1981, which furthered rail deregulation for eastern roads. Because of the high rate of inflation in 1980–82 (averaging 10 per cent for the three years) the favorable results of the Staggers Act were not immediately apparent. But within a few years American railroads enjoyed a major increase in freight traffic and also a marked growth in income.

A number of railroad mergers quickly followed the passage of the Staggers Act. On November 1, 1980, Hays T. Watkins, president of the Chessie System, created CSX as a giant holding company for future railroad mergers. Two years later the Seaboard Coast Line Railroad merged with the Louisville & Nashville to become the Seaboard Coast Line Industries, often known as the Family Lines. CSX controlled both the Chessie and the Seaboard Family Lines. Newspaper reporters were soon being told by Watkins that the C stood for Chessie, the S for Seaboard, and that the X meant that the system was much larger than one plus one. Only three years after its creation, CSX had assets of over $7.5 billion, 70,000 employees, and 27,000 miles of line in 22 states. CSX also had interests in trucking, barge lines, and warehousing. Watkins said, "CSX is no longer solely a railroad company. We must be a market-driven transportation company or be driven out of the market."

Also in November 1980, the 4,000-mile Frisco, with lines from Kansas City and St. Louis south to Dallas and Mobile, merged with the Burlington Northern, creating a 27,000-mile network. Another large system, the Norfolk Southern Corporation, appeared in June 1982 with the merger of the Norfolk & Western Railway and the Southern Railway. The Norfolk Southern in 1986 had assets of nearly $10 billion, annual revenues of $4 billion, and 38,000 employees. Its 17,000-mile network extended from Norfolk to Omaha, and from Chicago, Detroit, and Buffalo south to New Orleans, Mobile, and Jacksonville.

By 1982 the Union Pacific had acquired the 1,100-mile Western Pacific and the much larger Missouri Pacific with lines in eleven different states, creating a combined system of 21,000 miles. Later in 1988 the Union Pacific would also take over the "KATY" or Missouri-Kansas-Texas Railroad. The Southern Pacific had held a major portion of the capital stock of the St. Louis Southwestern Railway (Cotton Belt Route) since the depression 1930s, and was in full control of the Cotton Belt line by the mid-1980s.

None of these newly merged and enlarged systems (Conrail, Burlington Northern, CSX, Norfolk Southern, or Union Pacific) retained their original massive mileage for long. Most mergers had parallel or duplicate lines that were really surplus. Unprofitable branch lines could be sold, leased, or abandoned. Poorly maintained routes or those with steep grades were also candidates for elimination. Within only three or four years, hundreds and hundreds of surplus route miles would be disposed of in some way. Some of the cutbacks in mileage were quite drastic. The 9,500-mile Illinois Central Gulf of 1972 was reduced by 4,700 miles by 1985, and further still to 3,000 miles by 1990. The entire IC route

Harley O. Staggers. A veteran West Virginia Congressman, Staggers sponsored the Staggers Rail Act which greatly reduced the regulatory powers of the ICC. (Courtesy, West Virginia State Archives.)

Hays T. Watkins. In the 1970s and 1980s Watkins created the Chessie System, and later CSX, out of several large Class I railroads. (Courtesy, CSX Transportation.)

L. Stanley Crane. In the 1980s Crane turned Conrail from red ink to dividends, and in 1987 sold it to private investors. (Courtesy, Norfolk Southern Corporation.)

W. Graham Claytor, Jr. For more than a decade Claytor improved Amtrak service, and pushed its Revenue to Expense Ratio up to 79 per cent. From *The Southern Railway: Road of the Innovators* by Burke Davis. Copyright © 1985 by the University of North Carolina Press. Used by permission of the publisher.

Union Pacific unit coal train being loaded from mine tower without having to stop. (Courtesy, Union Pacific Railroad.)

across Iowa, the original Gulf, Mobile & Ohio, as well as many branch lines, had all been dropped. Hays Watkins reduced his CSX from nearly 28,000 miles in 1980 to 24,000 miles in 1985, and to 19,000 miles by 1990. Many of these lines that were dropped from recently merged systems as being surplus or expendable seemed essential to the shippers in the local areas. Those local shipping interests frequently took over and operated the lines as regional railroads, sometimes with modest subsidies from local or state governments.

The mergers materially reduced the number of railroads in op-

CSX double stacked container freight train. (Courtesy, CSX Transportation.)

eration. As the ICC changed the definition of Class I railroads, the number of Class I lines was also reduced. For the first half of the twentieth century the ICC definition of a Class I railroad was any line with operating revenue of more than $1,000,000 a year; Class II railroads had annual revenues of from $100,000 to $1,000,000; and Class III lines were under $100,000. In 1955 there were 126 Class I railroads in the nation. In 1956 Class I lines were required to have at least $3,000,000 of annual revenue, and by 1963 the number of such railroads had dropped to 102. The revenue needed for Class I status was raised in 1965 to $5,000,000 and soon only 76 lines were so listed.

The $10,000,000 figure effective in 1976 for Class I lines reduced the number to only 52, and the $50,000,000 requirement effective in 1978 cut Class I roads to only 41. Clearly, the revenue needed for Class I status was rising much faster than inflation. The definition of $88,000,000 of revenue needed in 1987 cut Class I lines to only 16. In addition there were 27 Regional railroads and 457 lines defined as Local (including Switching and Terminal lines). By 1995 the numbers were: Class I, 11; Regional, 30; and Local railroads, 500. The labor force required for the Regional and Local roads in the early 1990s came to only half a worker per mile of track.

The financial gains that many hoped would accompany the Staggers Act did not appear in the early 1980s. The inflation which averaged 10 per cent a year in 1980, 1981, and 1982 was not conducive to increased productivity and profit. Nor was the sharp, but short, depression of 1982 of any help. Good gains in revenue ton-mileage had appeared in the late 1970s, but the figure of 919 billion in 1980 was to be exceeded only once (921 in 1984) prior to 1987. The 1987 figure of 944 billion ton-miles increased annually for the next eight years—996 in 1988, 1,034 in 1990, 1,067 in 1992, and 1,305 in 1995. Most of the gains in these years were in the western rather than the eastern states. Somewhat surprisingly the operating revenue of Class I railroads changed very little between 1980 and 1995, averaging $28.5 billion in the early 1980s, $27.2 billion in the late 1980s, and $29.3 billion in the early 1990s. In the same years the average ton-mile freight rate fell from 2.9 cents in 1980 to 2.4 cents in 1995 in current dollars. The decline was even greater in constant dollars because of continuing inflation.

In the years after 1980, most Class I railroads managed to reduce their operating expenses substantially. Between 1980 and 1995 the average train load increased 29 per cent, the average length of train haul 36 per cent, and the average freight car capacity 14 per cent. In the same fifteen years, freight loss and damage claims dropped from $285,000,000 a year to $102,000,000 a year. Other expenses also declined in the decade and a half. In 1980 total labor costs (including payroll taxes) had taken 52 cents out of each revenue dollar. In 1995 the total labor cost per revenue dollar was below 41 cents. In the same period, fuel costs dropped from 11.5 cents to 6.5 cents per revenue dollar.

As a result of the lower yearly expenses, the rate of return on shareholder's equity took a decided upturn in the 1980s and 1990s as compared to the 1960s and 1970s. Between 1960 and 1979 the average rate of return was only 2.3 per cent for the twenty years (in only four years was the return above 3 per cent). In the years between 1980 and 1995 the average rate of return on the investment was 7.4 per cent (in five of those years the return was 9 per cent or higher). Quite clearly the Staggers Act had returned prosperity to American railroads. The average rate of return for the four years 1992–95 was 9.4 percent. Many observers felt this figure was still shy of "revenue adequacy" (a rate of return equal to the cost of capital) which the Staggers Act was supposed to provide. In the years since 1980, Class I railroads reduced their debt. In 1980 their capital structure was 40 per cent debt and 60 per cent equity. By 1995 the debt was 20 per cent and the equity up to 74 per cent.

A high point for American railroads in the 1980s was the success Conrail achieved by being sold to private interests in 1987. When Ronald Reagan became president in 1981, he stressed a need to cut federal spending sharply, and he believed that Conrail should be disposed of quickly, if possible. Secretary of Transportation Drew Lewis agreed with Reagan concerning the future of Conrail. On January 1, 1981, three weeks before Reagan's inauguration, L. Stanley Crane became chairman and top executive officer of Conrail. Crane was an engineer-chemist who had worked his way up through the ranks of the technical-engineering side of the Southern Railway from 1937 to 1976, becoming president and chairman of the railroad from 1976 to 1980 and a leader in the modernization of the Southern Railway after World War II. Crane believed the previous management of Conrail, using billions of federal dollars, had done a good job of restoring the Conrail property. He hoped to streamline the entire system and make it operate more effectively. It soon became evident that President Reagan and Lewis were willing to give him his chance to do so.

Crane was greatly aided by the 1980 Staggers Act, since it allowed him to raise certain freight rates without going to the ICC. The Northeast Rail Services Act of 1981 also freed Crane from providing commuter passenger services, which had never made a profit for Conrail. However, Conrail was still losing money, and Crane managed to obtain one final federal line of credit for $300 million. Crane once said: "In the railroad business . . . you've got to market in places where other people are not doing the job." He went after new traffic, the major areas of growth being coal, ore, grain, finished automobiles, and intermodal freight. During the early 1980s the intermodal traffic increased over 50 per cent, much of it moving in double-stacked container service.

By 1982 Conrail was making a small profit, which climbed to $313 million in 1983 and even more the next year. Crane not only made money by serving new markets, but saved more with widespread cost cutting. He reduced the system by over 4,000 route-miles with only minimal effect to either the volume of traffic or the total revenue. Within two or three years he had also decreased the total cost of Conrail labor from 54 to 45 cents of each revenue dollar. Crane managed to cut back the average crew size to 3.1 members per freight train, one of the lowest in the industry. In the same years he continued to make capital investments to improve his plant, laying down high-speed welded rail on his main line tracks, investing in labor-saving mechanized maintenance equipment, and buying new high-tech diesel locomotives.

The Northeast Rail Services Act of 1981 had stated that Conrail should be sold to private ownership in 1984. A number of railroads, including Norfolk Southern, Santa Fe, and CSX, had expressed an early interest in acquiring Conrail. Elizabeth Dole, the Secretary of Transportation who had succeeded Drew Lewis in 1983, selected Norfolk Southern with its offer of $1.2 billion as the winner in the bidding contest. Although Norfolk Southern later increased its bid to $1.9 billion, Congress failed to approve the sale of Conrail to Norfolk Southern. There was a growing belief, strongly supported by Crane, that Conrail should be sold by the government with a public offering of the common stock of the railroad. Conrail was sold to the public with 58 million shares sold at $28 a share on the New York Stock Exchange on March 26, 1987. The following day Conrail stock opened at $31.50 a share. Crane remained with Conrail until 1988, as the public-owned line continued to prosper.

In the 1980s, Amtrak was a much different story than Conrail. When Congress set up Amtrak in 1970–71, it did not expect that at some future date the railroads might resume passenger service as a private for-profit venture. However, much of the general public was not aware that all the industrialized nations, except for the United States and parts of Canada, had government-owned or socialized railroad systems. But Congress did hope, and expect, that eventually Amtrak ticket revenue would pay for much, or most, of the expenses of its operations. In 1980, Amtrak was paying less than half of its way with ticket receipts. In July 1982, W. Graham Claytor, Jr. (1912–1994), a Washington, D.C., lawyer and an effective past president of the Southern Railway from 1967 to 1976, succeeded Alan Boyd as chairman and president of Amtrak. Claytor knew his way around Washington since he had been Secretary of the Navy from 1977 to 1979 in the Carter Administration. But now, Claytor was serving under Ronald Reagan, a president eager to trim the federal budget wherever he could. A Reagan adviser joked that it would be cheaper just to give every Amtrak patron "a bus ticket, newspaper, and a three-martini lunch."

But there was solid support for Amtrak in both houses of Congress. Many members of Congress received hundreds of letters supporting Amtrak. Most representatives and senators agreed with Robert C. Byrd, veteran senator from West Virginia, that Amtrak must certainly serve the senator's own state. Claytor obtained many new passenger cars in the 1980s from the Budd Company of Philadelphia, and many older cars were rebuilt at the Amtrak Shops at Beech Grove, Indiana. Throughout his eleven years with Amtrak, Claytor never felt there were sufficient funds available for the capital needs of the system. Money was so tight during the eighties that disposable plastic rather than china dinnerware was often used in the dining cars. During the 1980s, Claytor did achieve longer work days for Amtrak train crews, nearly gaining a 40-hour work week. He also improved the speed and service on the Metroliners. Metroliner schedules were so fast that many travelers preferred Amtrak over the airlines between Washington, D.C., and New York City. Claytor's major accomplishment was in cost cutting and general economy of operation. By 1986 he had pushed the Revenue to Expense Ratio up to 62 per cent, and by the early 1990s it had risen to 79 per cent. He was also a major sponsor of the multi-million dollar restoration of the Washington

Union Station, which reopened as the Visitor Center in September 1988.

Even with Claytor's excellent management, Amtrak did not change much during the eleven years of his tenure. It continued to have a roster of 24,000 employees, serving 1,900 cars pulled by about 400 locomotives over a 25,000-mile network with 500 stations in 44 or 45 of the contiguous 48 states. Amtrak trains did not begin to match the service, speed, or on-time performance of the trains of the post–World War II and early 1950s. Most Amtrak passengers were families on vacation, the elderly and the retired, travelers who did not wish to fly, or people who just liked to ride trains. The 5.3 billion passenger-miles of Amtrak travel in 1987 amounted to only 2 per cent of the airliner travel that year.

In 1993, Thomas M. Downs, former director of the Urban Mass Transit Administration, was selected to replace the retiring W. Graham Claytor, Jr. When Downs had to face cuts in federal funding for Amtrak during the Bill Clinton presidency, some states volunteered subsidies to keep certain Amtrak trains in service. Downs also put in service new mail express cars to increase Amtrak revenue. In 1995–96 the Republican House of Representatives voted heavily to keep Amtrak alive to the end of the century. But schedules had to be trimmed in 1995–96, and several routes were changed from daily to triweekly service. Later, in the fall of 1996, several major routes (Denver to Seattle, Salt Lake City to Los Angeles, and St. Louis to San Antonio) were eliminated, with daily service resumed on most remaining routes. Amtrak officials decided: "You either have to be daily in a market or not in a market." When Amtrak celebrated its Silver Anniversary in the spring of 1996, not many people were willing to predict its future.

American railroads continued to downsize in the 1980s and 1990s. A few railroaders called it "re-engineering" instead of downsizing, and felt the reductions were excessive. But the cutbacks did continue. Total mileage for all classes of railroads was 179,000 miles in 1980, 175,000 in 1990, and 170,000 in 1995. For Class I lines the mileage was 164,000 miles (40 lines) in 1980, 133,000 miles (14 lines) in 1990, and 125,000 miles (11 lines) in 1995. Class I employment was 458,000 in 1980, but dropped sharply to 216,000 in 1990 and to only 186,000 in 1995, a figure well under half that of 1980. Annual wages for Class I railroads increased rapidly in the decade and a half, climbing from $24,698 in 1980 to $34,991 in 1985, to $39,987 in 1990, and to $48,188 in 1995. These increases were well above the inflation rate for those years. A freight engineer in 1995 with a good run might well earn $65,000 a year or more. Since the Class I labor force had dropped more than half, and average wages had not quite doubled in the same period, the total compensation had dropped. Total compensation was $11.3 billion in 1980, and $9.0 billion in 1995.

While wages had climbed in these years, rail management had finally gained a modest change in the definition of a "day's work" for freight crews. Since the years of federal control in World War I a day's work for a freight crew had been 100 miles. Since through freights could average 20 to 30 miles an hour, and a typical run might be well over 100 miles, a day's pay could often be earned in well under eight hours. After long months of negotiations in the mid-1980s, a national mediation board agreed that starting in 1986 a day's work would be raised to 104 miles. Later increases were: 106 miles in 1987, 108 miles in 1988, 114 miles in 1991, 118 miles in 1992, 122 miles in 1993, and 130 miles in 1995. In the mid-1990s, several freight train accidents were blamed on overtired engine crews who had spent too many hours of service in a 24-hour period. As of 1996 it did not seem probable that a freight crew's workday would be raised above the 130-mile figure.

Railroad unions remained strong in the 1980s and 1990s, and these changes in the definition of a "day's work" came only after long, intensive, and often bitter negotiations. But railroaders, from section man to top manager, continue to be proud of their occupation, and their basic devotion to their company and industry remains strong, as it has for the last century and a half.

Modest technical improvements in rail freight service continued to appear in the 1980s and early 1990s. Unit coal train traffic still climbed, and in 1995 coal traffic on Class I lines accounted for over 40 per cent of the tonnage and nearly 22 per cent of the revenue. Newly built aluminum coal gondolas could carry 122 tons as compared to 100 tons for the older cars. Lighter-weight flatcars were also used to carry double-stacked containers. Intermodal container traffic nearly doubled between 1988 and 1995, growing much faster than the earlier piggyback service. Hays Watkins's CSX containers moved to and from his system to ships in the Gulf, Atlantic, and Pacific waters. The Sea-Land Service segment of CSX in 1994 operated a fleet of 93 container ships and 188,000 containers for its U.S. and foreign trade. The lighter freight cars along with more efficient and powerful diesels permitted the revenue ton-miles per gallon of fuel oil to increase from 235 in 1980 to 375 in 1995. By the early 1990s, new diesel units with two six-wheel trucks were available with 6,000 horsepower, and could be built with either direct or alternating current. The two major producers of diesels were General Motors' Electro-Motive Division and General Electric's Transportation System. During these years the use of ribbon or welded continuous rail also became more common. Some lines also started to experiment with concrete rather than wooden crossties in the 1980s and 1990s.

The years of downsizing, cost cutting, and technical improvements in freight service together made possible three landmark events in 1995. That year American railroad freight traffic reached a record high of 1,305 billion ton-miles, nearly twice the figure for 1944, the top year in World War II. The share of intercity freight moving by railroad had dropped below 36 per cent in the mid-1980s. Then it had started to climb—37.0 per cent in 1988, 37.6 per cent in 1990, 38.1 per cent in 1993, 39.1 per cent in 1994, and

40.6 per cent in 1995. The growing efficiency in rail operation in the 1980s had seen the annual freight service per worker double in eight years from 2.1 million ton-miles in 1980 to 4.2 million ton-miles in 1988 and then soar to 7 million ton-miles for the year 1995.

A major change in these years was the slow disappearance of the caboose. A study by the Association of American Railroads in 1987 indicated that using an EOTD (end-of-train-device) was as safe as a caboose and would save 57 cents per freight train mile. These EOTDs not only protected the rear end of the train with a red light, but also provided information to the conductor and engineer in the diesel locomotive. Crew members informally called the device FRED (flashing-red-end-device). By the early 1990s most freight trains carried a FRED on the rear car. The caboose was going the way of the steam engine, the train-order-loop, and the small town depot. A few lines still used cabooses occasionally, but most railroads considered them surplus. Railroad buffs could purchase a caboose for their rural retreat for $5,000 from more than one railroad.

In the mid-1990s, several more important mergers occurred. In July 1995, the ICC approved, with a 4–0 vote, the merger of the 22,000-mile Burlington Northern with the 8,000-mile Santa Fe. The merged road, with a combined investment of $10 billion and revenues of over $7 billion, was to be called the Burlington Northern Santa Fe Corporation. Even earlier, in March 1995, Drew Lewis's 16,000-mile Union Pacific, with ICC approval, had acquired 70 per cent of the stock of the 4,000-mile Chicago & North Western. The C&NW, way back in the 1860s, had carried supplies to the Missouri River as the Union Pacific built its line across Nebraska, and for decades the C&NW had been the UP connection east to Chicago. In August 1995, the Union Pacific announced it intended to buy its rival, the 13,000-mile Southern Pacific. There was considerable opposition to the UP plan from Washington, and from both shippers and other railroads.

Before the decision was made on the UP-SP merger, the Interstate Commerce Commission itself was eliminated by votes in both Houses of Congress, and its elimination was approved by Bill Clinton in December 1995. The ICC, the oldest federal regulatory agency in the nation, was replaced on January 2, 1996, by a new, three-member Surface Transportation Board, within the Department of Transportation. The Surface Transportation Board took over the now reduced (since the Staggers Act) control over the railroads previously held by the ICC. On July 3, 1996, the Surface Transportation Board with a 3–0 vote approved the Union Pacific-Southern Pacific merger. The new UP-SP system with about 35,000 miles of track in twenty-five states was by far the largest rail system in the nation. In less than two years, five lines (BN, SF, UP, C&NW, and SP) had become two great systems.

As of mid-1996 the number of Class I railroads had been reduced to only nine (in order of length): UP-SP, BN-SF, CSX, Norfolk Southern, Conrail, Soo, Kansas City Southern, Illinois Central, and Grand Trunk Western. In mid-October 1996, CSX announced that it planned to buy Conrail for about $8 billion to become the third largest railroad (after UP-SP and BN-SF). In the winter of 1996–97, the Norfolk Southern tried to outbid CSX for the control of Conrail. If either the CSX-Conrail merger or the Norfolk Southern-Conrail merger was to be approved later, these giant roads, two in the West and one in the East, would account together for nearly 80 per cent of the line mileage, revenue ton-mileage, and employees of the Class I railroads in the nation.

In the early 1920s, as the railroads were recovering from 27 months of federal control, there had been a strong movement to merge the 200-plus Class I railroads into nineteen or twenty strong regional systems. This merger idea was not popular with the railroad industry in the 1920s or 1930s. But the dozens of mergers in the last few decades have remade the railroad map of the nation.

* * * * *

American railroads have come far in the 170 years since William Strickland (c. 1787–1854) tried to convince his fellow Philadelphians that railroads would surely supersede canals. Soon early railroads built by Baltimore, Charleston, and Albany were serving their local areas. By the 1850s, dozens of new roads were being constructed and a 30,000-mile rail network had reached the Mississippi and the western frontier line. Northern lines helped the Union to victory in the Civil War. As farmers, shippers, and consumers endured rate discrimination and financial corruption in the years of Gould and Vanderbilt, they seemed momentarily to agree with Charles Francis Adams, who had said it was "useless for men to stand in the way of steam engines." The railroads hastened the settlement of the trans-Mississippi frontier, and helped transform an agrarian-based, second-rate industrial society into the complex industrial nation we know today. Stiff federal regulations of railroads passed in the Progressive Era have continued through most of the twentieth century. Since the 1920s, the competition of new forms of transportation has caused a serious decline in rail traffic, especially in passenger service. However, the inherent economy of the flanged wheel running on a steel rail is so great that American railroads are still very viable today, and should be for many future decades.

Important Dates

1826   Gridley Bryant uses his broad-gauge tramway to haul granite for the Bunker Hill Monument

1828   John Quincy Adams turns the first dirt for the Chesapeake and Ohio Canal, July 4

Charles Carroll lays the first stone for the Baltimore & Ohio Railroad, July 4

1830   Peter Cooper's locomotive "Tom Thumb" used on the Baltimore & Ohio

First scheduled steam railroad train service in America, by "Best Friend of Charleston," at Charleston, South Carolina

At least 10,000 miles of canal actively projected and 1,277 miles completed

Twenty-three miles of railroad in operation in the United States

1831   Locomotive "DeWitt Clinton" pulls the first steam train in New York from Albany to Schenectady

First U.S. mail carried by rail

First important railroad periodical, *American Railroad Journal*, founded

1833   Andrew Jackson, first President to ride on a railroad, travels between Ellicott's Mills and Baltimore, Maryland

1834   Pennsylvania's 395-mile Main Line (combined rail, inclined-plane, and canal route) completed

1835   Boston to Providence rail line opened

Railroad completed to Washington, D. C., from Baltimore

1837   American-type locomotive (4-4-0) planned and built

First sleeping car (a crudely remodeled day coach) service

1839   Long-distance railway express service started by William F. Harnden

Important Dates

Establishment of first state railroad commission, in Rhode Island

1840   Nearly 3,000 miles of railroad and 3,300 miles of canal in the United States

1841   Rail line completed from Boston to the Hudson River (Albany)

1842   Rail service opened between Albany and Buffalo

1848   Rail service opened between Cincinnati on Ohio River and Sandusky on Lake Erie

        "Pioneer," Chicago's first locomotive, placed in operation

        Rail route completed from Boston to New York City

1850   Millard Fillmore signs the first railroad land-grant act, aiding the Illinois Central and the Mobile & Ohio

        Railroad mileage in the United States up to 9,000 miles

1851   Great celebration marks completion of the Erie Railroad to Dunkirk on Lake Erie

        Telegraph used for dispatching trains

1852   Philadelphia and Pittsburgh linked by rail (including inclined planes)

        Baltimore & Ohio completed to Wheeling on the Ohio River

1853   All-rail route opened from the East to Chicago

1854   Chicago and St. Louis connected by rail

1856   Railroad bridge across Mississippi completed at Davenport, Iowa

        Illinois Central, world's longest railroad (705 miles), completed

1857   Southern rail route from Charleston to Memphis put into operation

1859   First Pullman sleeping car, built by George M. Pullman, makes
         initial run

1860   Chicago, served by eleven railroads, becomes a major rail
         center

         Nation has rail network of 30,000 miles

1861–65 Railroads are important to both sides in the Civil War, and the
         northern roads make a major contribution to the ultimate
         Union victory

         During the war the combined freight tonnage of the Erie and
         the New York Central railroads for the first time exceeds that
         of the Erie Canal

1862   "Andrews Raid," in which Union soldiers stole the locomotive
         "General"

         Abraham Lincoln signs the first Pacific Railway bill

         Experimental postal car for the sorting of mail en route put
         into service between Hannibal and St. Joseph, Missouri

1863   Ground broken for the Union Pacific Railroad at Omaha and
         the Central Pacific at Sacramento

         Brotherhood of Locomotive Engineers (oldest of present rail-
         road unions) organized

1865   First oil tank car placed in service

         Manual block system of train control developed by Ashbel
         Welch

         First domestic steel rails are produced, but acceptance of them
         is slow

1866   "Co-operative" fast-freight lines appear

1867   First patent issued for a railroad refrigerator car

         John D. Rockefeller and Henry M. Flagler begin bargaining
         for rebates on shipments of oil

Important Dates

Joseph G. McCoy ships first longhorns out of Abilene, Kansas, the first major cowtown

"Erie War" started, with Drew, Gould, and Fisk of the Erie opposing Commodore Vanderbilt of the New York Central

1868 Eli H. Janney patents his automatic coupler

1869 George Westinghouse applies for a patent for his air brake

Driving of the Golden Spike marks completion of first transcontinental rail route, May 10

1870 National railroad system has increased to 53,000 miles

1871 First narrow-gauge line opened (near Denver, Colorado)

Last railroad land grant made by the federal government

Granger railroad regulation appears, with legislation in Illinois and Minnesota

1874 Windom Committee (U.S. Senate) presents a report critical of railroad abuses

Granger legislation appears in Iowa and Wisconsin

1876 U.S. Supreme Court upholds Granger Laws in *Munn* v. *Illinois*

1877 Serious railroad labor trouble starts on the Baltimore & Ohio and spreads across the country during July

1880 National rail network has grown to 93,000 miles

1881 Steam heat for passenger equipment introduced

1883 Completion of Northern Pacific, second railroad to the Pacific

Standard time adopted by the railroads of the nation

1885 J. P. Morgan brings peace to a railroad rate war between the New York Central and the Pennsylvania

1886   Cullom Committee (U.S. Senate) reports on railroad abuses

Southern railroads shift from five-foot to standard-gauge track

Supreme Court holds that a state cannot regulate interstate commerce

1887   Interstate Commerce Act passed by Congress provides for first federal regulation of railroads

Introduction of vestibule passenger train

New construction hits an all-time high of 12,878 miles for the year

First trains fully equipped with electric lights appear

1890   National rail network has expanded to 164,000 miles

Great majority of all mileage is laid with steel rails

1893   Railroad Safety Appliance Act requires that all trains be equipped with automatic couplers and air brakes

New York Central locomotive "No. 999" attains a speed of 112.5 miles per hour

1894   Pullman Strike, May to July

1895   First electrified locomotive train service in America

1900   National rail network stands at 193,000 miles

1903   Elkins Act passed by Congress attempts to strengthen the prohibition of railroad rebates

1904   U.S. Supreme Court orders dissolution of the Northern Securities Company

1906   Hepburn Act passed by Congress greatly increases the powers of the Interstate Commerce Commission

All-steel passenger cars placed in regular service

1907   Union Station in Washington, D.C., built

Important Dates

1910   Mann-Elkins Act passed by Congress further extends the jurisdiction of the ICC

National rail network at 240,000 miles

Pennsylvania Station finished in New York City

1914   World War I brings increase in rail traffic

First use of radio in railroad communications

1916   Adamson Act provides an eight-hour day for operating personnel

National rail mileage at an all-time peak of 254,000 miles

Railroads of the nation carry 77 per cent of the intercity freight traffic and 98 per cent of the intercity passenger business

Federal government provides new grants-in-aid for highway construction

1917   National shortage of freight cars rises to 158,000 cars

Government takes over operation of railroads on December 28

1918   Post Office Department starts air-mail service

1920   Esch-Cummins Act, or Transportation Act of 1920, helps in the return of the railroads to private management

Railroads returned to private management on March 1

Railroad employment exceeds 2,000,000

Car Service Division set up within the Association of American Railroads

National rail system drops to 253,000 miles

1923   Railroads embark upon a comprehensive program of improvement and increased efficiency

Creation of first Shippers Advisory Board

1925   First diesel-electric locomotive in regular switching service

American Railroads

1926	Commercial airlines carry 5,800 passengers in the year
1927	Centralized Traffic Control (CTC) has first installation
	First experiments with air-conditioned passenger cars
1929	Greyhound bus system starts service
1930	Air-conditioned passenger cars appear in regular service
	Rail passenger service from 1930 through 1959 operated with annual deficits except for the years 1942–45
	National rail system drops to 249,000 miles
1933	Emergency Railroad Transportation Act passed to aid a rail industry hurt by the depression of the thirties
1934	Diesel locomotives first used in passenger service
	Introduction of lightweight streamlined passenger trains by the Burlington and the Union Pacific
1939	World War II brings increases in rail traffic
1940	Transportation Act of 1940
	National rail network drops to 233,000 miles
1941–45	During World War II, the railroads meet the transport needs of the war without government management
1941	Diesel locomotives first used in freight service
1945	Vista-dome passenger equipment introduced
1950	National rail network drops to 224,000 miles
1954	First use of television in railroad communications
	Piggyback freight service offered by several railroads
1956	Economy slumbercoaches first used

Important Dates

1957   Passenger movement by air exceeds that by rail

1958   Transportation Act of 1958

1959   National rail network drops to 220,000 miles

   Rail employment declines to just over 800,000

   Railroads of the nation carry only 44 per cent of the intercity freight traffic and only 28 per cent of the intercity commercial passenger traffic

1960   Unit coal train operations begin

1964   U.S. Supreme Court decides most freight train and yard firemen should be eliminated

1965   Piggyback freight service exceeds 1,000,000 carloadings

1966   Pennsylvania-New York Central merger approved by ICC

1970   Penn Central files for bankruptcy

1971   Federal operation of Amtrak passenger service begins

1976   Conrail begins freight service in eastern United States

1980   Staggers Rail Act reduces ICC regulation of railroads

   National rail network drops to 179,000 miles with 512,000 employees

1982   Conrail in the black with a small profit

1986   Amtrak revenue reaches 62 per cent of yearly expenses

   Rail management makes an agreement with labor to redefine pay scale for freight crews

1987   Conrail common stock sold to general public

1990   Mergers and new definition of Class I lines reduce Class I railroads to only fourteen in number

1995 Congressional action, approved by President Clinton, eliminates ICC as a regulatory agency

National rail network declines to 170,000 miles with 209,000 employees and only eleven Class I lines

Rail freight service is 1,305 billion ton-miles, nearly twice the figure for 1944, the top year in World War II

Suggested Reading

Railroad literature is both varied and extensive. Americans started to write about their railroads even before they built them, and they have never stopped writing. In 1957 the Lexington Group, an informal organization of railroad historians, prepared a checklist of secondary works in railway history. Admittedly far from complete, the list still included more than two thousand different titles. Not long after, the Library of Congress was acquiring new books on railroads and railroading at a rate of over two hundred titles per year.

GENERAL WORKS ON RAILROAD AND TRANSPORTATION HISTORY

Though out of date, John L. Ringwalt, *Development of Transportation Systems in the United States* (1888), is still useful. The book suffers from faulty organization, but a vast variety of information on early American transportation is included. For the development of railroads in America, a popular but excellent survey is Stewart H. Holbrook, *The Story of American Railroads* (1947). *Hear the Train Blow* (1952), by Lucius Beebe and Charles Clegg, is a delightful pictorial epic of the railroad in nineteenth-century America. Valuable, though perhaps somewhat biased since it is written by a professional railroader, is the book by Robert S. Henry, *This Fascinating Railroad Business* (1942). For a detailed pictorial history of the same subject see Roderick Craib, *A Picture History of U.S. Transportation* (1958).

Two books by Roger Burlingame, *March of the Iron Men* (1938) and *Engines of Democracy* (1940), stress the role of invention and its effect on the American economic environment. The place of business in American history is discussed in Thomas C. Cochran and William Miller, *The Age of Enterprise* (1942). A wealth of stories, legends, and tall tales of American railroaders and railroads is given in B. A. Botkin and Alvin F. Harlow (eds.), *A Treasury of Railroad Folklore* (1953). Edward Hungerford, *Locomotives on Parade* (1940), and Edwin P. Alexander, *Iron Horses: American Locomotives, 1829–1900* (1941), are both richly illustrated histories of the steam locomotive in America. Howard Fleming, *Narrow Gauge Railroads in America* (1949), is a new edition of a work which when first published in 1875 helped sell the narrow-gauge railroad to western America. A popular history of the express business is found in Alvin F. Harlow, *Old Waybills* (1934).

There are several good regional studies of American transportation. The best is Edward C. Kirkland, *Men, Cities and Transportation: A Study in New England History, 1820–1900* (2 vols., 1948). This is definitive on the railroads of New England. William F. Gephart, *Transportation and Industrial Development in the Middle West* (1909), covers the history of highways, canals, and railroads in nineteenth-century Ohio. Robert E. Riegel, *The Story of Western Railroads* (1926), is an excellent general account of western rail development.

Several books treat the financial and political aspects of American railroading. William Z. Ripley, *Railroads: Finance and Organization* (1915), gives a thorough review of the financial structure of all the rail system just as it was achieving its full physical maturity. *Railroads: Rates and Regulation* (1912), by the same author, discusses the problem of rates as their regulation was becoming nearly complete. Ripley is also the editor of *Railway Problems* (1913), a valuable collection of railway source reprints. An excellent treatment of local, state, and national aid in the building of the American rail system is Frederick A. Cleveland and Fred W. Powell, *Railroad Promotion and Capitalization in the United States* (1909). The effect of the railroad on the legal structure of Wisconsin in the nineteenth century is presented in Robert S. Hunt, *Law and Locomotives* (1958).

Two recent works, Robert L. Frey (ed.), *Railroads in the Nineteenth Century* (1988), and Keith L. Bryant, Jr. (ed.), *Railroads in the Age of Regulation, 1900–1980* (1988), review the history and biography of many railroads and rail officials in some detail. Alfred D. Chandler, Jr., *The Railroads, The Nation's First Big Business* (1965), makes the point that early railways promoted economic change and began modern corporate finance. In contrast, Robert W. Fogel, *Railroads and American Economic Growth* (1964), presents a thesis that railroads were not very important in American economic growth in the nineteenth century. A lively and richly illustrated review of the nation's railways is Oliver Jensen, *The American Heritage History of Railroads in America* (1975). John F. Stover, *The Life and Decline of the American Railroad* (1970), outlines the decline in service and profitability of railroads since World War II. A later work by Albro Martin, *Railroads Triumphant* (1992), overstates their recent rebirth and growth. Three definitive works by John H. White, Jr., are *American Locomotives, An Engineering History, 1830–1880* (1968), *The American Railroad Passenger Car* (1978), and *The American Railroad Freight Car* (1993). With hundreds of pictures and drawings, and over 1,800 pages of text, the three volumes are the last word on early steam power and rolling stock. A detailed review of the long history of Pullman sleeping cars is found in Peter T. Maiken, *Night Trains* (1989). Pictures and floor plans of some 300 depots (of the 80,000 stations of the 1940s) are found in

American Railroads

The Country Railroad Stations in America (1978) by H. Roger Grant and Charles W. Bohi. George W. Hilton, *American Narrow Gauge Railroads* (1990), is an exciting review of the history of the 12,000 miles of narrow-gauge line built a century ago by over 350 individual companies. Two excellent monthly magazines for both railroad historians and rail buffs are *Trains* and *Railway Age*. *Railroad Facts*, published yearly by the Association of American Railroads, is a useful source for statistics for Class I railroads.

HISTORIES OF SPECIFIC RAILROAD COMPANIES

Books on specific railroads tend to offer more in quantity than quality. Of the histories of eastern lines, one of the best is Edward Hungerford, *The Story of the Baltimore and Ohio Railroad, 1827–1927* (2 vols, 1928). Quite satisfactory is the popular history, *The Road of the Century: The Story of the New York Central* (1947), by Alvin F. Harlow. H. W. Schotter, *The Growth and Development of the Pennsylvania Railroad Company* (1927), and George H. Burgess and Miles C. Kennedy, *Centennial History of the Pennsylvania Railroad Company* (1949), are both uncritical "company" histories. Edward H. Mott, *Between the Ocean and the Lakes: The Story of the Erie* (1901), is the standard study of the Erie in the nineteenth century. More lively is Edward Hungerford, *Men of Erie* (1949). Another popular history is Taylor Hampton, *The Nickel Plate Road* (1947). Several coal lines are presented in Jules I. Bogen, *The Anthracite Railroads: A Study in American Enterprise* (1927).

One of the best volumes on southern lines is Howard D. Dozier, *A History of the Atlantic Coast Line* (1920), reviewing the consolidation of more than a hundred short southern roads. A friendly and complete history of the Chesapeake and Ohio Railway is Charles W. Turner, *Chessie's Road* (1956). Thomas D. Clark, *The Beginning of the L. and N.* (1933), is an excellent, though brief, account for the pre–Civil War years. More complete is the "company" history, *The Louisville and Nashville Railroad, 1850–1942*

276

(1943), by Kincaid A. Herr. The construction of several short lines is fully traced in Thomas D. Clark, *A Pioneer Southern Railroad, New Orleans to Cairo* (1936).

Histories are also available for a number of lines west and south of Chicago. Best for the Illinois Central is Carlton J. Corliss, *Main Line of Mid-America* (1950). The utilization of the land grant by the Illinois Central is admirably covered in Paul W. Gates, *The Illinois Central and Its Colonization Work* (1934). Equally good for the land policy of the Burlington is *Burlington West* (1941), by Richard C. Overton. Overton is the author of another excellent book, *Gulf to the Rockies* (1953), the story of Burlington-built lines south of Denver in the years 1861–98. Several quite commonplace books have been written on America's first transcontinental line, and the best is probably Nelson Trottman, *History of the Union Pacific* (1925). The volume by the road's chief engineer, Grenville M. Dodge, *How We Built the Union Pacific* (1910), must be used with caution. Stuart Daggett, *Chapters on the History of the Southern Pacific* (1922), is a useful study of that road's extensions from Portland, Oregon, to El Paso. Adequate for the Santa Fe are James Marshall, *Santa Fe* (1945), and L. L. Waters, *Steel Rails to Santa Fe* (1950).

A number of additional railroad histories have appeared in the last three decades. A fully detailed and definitive history of one of the strongest "Granger" lines is Richard C. Overton, *Burlington Route: A History of the Burlington Lines* (1965). Another of the "Hill roads" is fully reviewed in *The Great Northern Railway: A History* (1988) by Ralph W. Hidy, Muriel E. Hidy, Roy V. Scott, and Don L. Hofsommer. The story of the first rail line to aim for the Pacific is told in two excellent volumes by Maury Klein, *Union Pacific, The Birth of a Railroad, 1862–1893* (1987), and *Union Pacific, The Rebirth, 1894–1969* (1989). The twentieth century history of the road that shared the 1869 Golden Spike celebration is *The Southern Pacific, 1901–1985* (1986) by Don L. Hofsommer. An earlier book by

Maury Klein, *History of the Louisville & Nashville Railroad* (1972), covers a major southern road. The history of one of the longest American railroads is found in Keith L. Bryant, Jr., *History of the Atchison, Topeka and Santa Fe Railway* (1974). The story of the first land grant railroad is John F. Stover, *History of the Illinois Central Railroad* (1975). Another book by John F. Stover is *History of the Baltimore and Ohio Railroad* (1987), the story of one of the first projected American railroads. Burke Davis, *The Southern Railway, Road of the Innovators* (1985), reviews a railroad made up of dozens of short southern lines of the nineteenth century. *The Nickel Plate Story* (1965), by John A. Rehor is the history of a late nineteenth century northern line that many observers considered to be an unnecessary railroad built only for its nuisance value. Robert F. Archer, *Lehigh Valley Railroad* (1977), is a pictorial history of an eastern line serving the anthracite coal fields of Pennsylvania. George W. Hilton, *Monon Route* (1978), is the best of several recent histories of the 500-mile "Hoosier Line." *The Corn Belt Route* (1984), by H. Roger Grant, is an excellent history of the Chicago Great Western Railroad. Robert G. Athearn, *Rebel of the Rockies* (1962), is a detailed history of the Denver and Rio Grande Western Railroad. M. C. Poor, *Denver South Park & Pacific* (1976), is a colorful history of several of Colorado's narrow-gauge lines.

BIOGRAPHY

A good place to find brief biographies of railroad leaders is the *Dictionary of American Biography*, edited by Allen Johnson Dumas Malone (22 vols., 1928–58). Thomas C. Cochran, in *Railroad Leaders, 1845–1890* (1953), gives an excellent review of the thinking of sixty executives of representative railroads. *The Age of the Moguls* (1953) by Stewart H. Holbrook is a fairminded and exciting treatment of many post–Civil War figures, including a number of important railroad men. William Miller (ed.), *Men in Business* (1952), consists of several scholarly studies of the American business world. Two volumes from the *Chronicles of America* by John Moody, *The Masters of Capital* (1919) and *The Railroad Builders* (1921), also include biographical material on the railroad magnates.

Of individual biographical studies, Archibald D. Turnbull, *John Stevens: An American Record* (1928), is adequate for one of the earliest railroad leaders. For the best biography of Cornelius Vanderbilt, see Wheaton J. Lane, *Commodore Vanderbilt* (1942). Julius Grodinsky, *Jay Gould: His Business Career, 1867–1892* (1957), provides much information for a major railroad buccaneer. W. A. Swanberg, *Jim Fisk: The Career of an Improbable Rascal* (1959), and Robert H. Fuller, *Jubilee Jim: The Life of Colonel James Fisk, Jr.* (1928), are both satisfactory for Fisk. Alfred D. Chandler, Jr., *Henry Varnum Poor: Business Editor, Analyst, and Reformer* (1956), is an admirable study of a major railroad figure. Henry G. Pearson, *An American Railroad Builder, John Murray Forbes* (1911), is still a useful biography of an easterner who built much middle western mileage. Lewis Corey, *The House of Morgan* (1930), is generally critical in tone. Somewhat more objective is Frederick Lewis Allen, *The Great Pierpont Morgan* (1949). M. W. Schlegel, *Ruler of the Reading: The Life of Franklin B. Gowen* (1947), is the study of both a man and his railroad.

Several volumes are also available for the builders of the western lines. For an objective treatment of the long career of Grenville M. Dodge, see J. R. Perkins, *Trails, Rails and War: The Life of Grenville Dodge* (1929). Still better reading, on the builders of the Central Pacific, can be found in Oscar Lewis, *The Big Four* (1938). An abundance of railroad material is included in E. P. Oberholtzer, *Jay Cooke, Financier of the Civil War* (2 vols., 1907). Also very useful is the more recent Henrietta Larson, *Jay Cooke, Private Banker* (1936). Henry Villard, *Memoirs of Henry Villard* (2 vols., 1904), should certainly be supplemented by James B. Hedges, *Henry Vil-*

lard and the Railways of the Northwest (1930). Joseph G. Pyle, *The Life of James J. Hill* (2 vols., 1917), is overly appreciative, and one might better use the briefer, more objective *James J. Hill: A Great Life in Brief* (1955), by Stewart H. Holbrook. For a detailed account of one of Hill's adversaries, see George Kennan, *E. H. Harriman* (2 vols., 1922).

A number of important biographies of railroad leaders have been written in the last thirty years. Maury Klein, *The Life and Legend of Jay Gould* (1986), is a definitive work providing needed reconsideration of the varied career of Gould. *The Great Persuader* (1969), by David Lavender, is a detailed biography of Collis P. Huntington, the most important of the builders of the Central Pacific. The life of Leland Stanford, another member of the "Big Four," is well presented in Norman E. Tutorow, *Leland Stanford: Man of Many Careers* (1971). A detailed study of one of the builders of the Union Pacific is found in Stanley P. Hirshson, *Grenville M. Dodge* (1967). Albro Martin, *James J. Hill and the Opening of the Northwest* (1976), is an excellent review of the life and work of the builder of the Great Northern Railway. Richard C. Overton, *Perkins / Budd* (1982), is a fascinating business biography of two Burlington Railroad top executives. *Bonds of Enterprise* (1984) by John L. Larson is the story of John Murrey Forbes, the Boston merchant who was a major sponsor of the Burlington in the midnineteenth century. James A. Ward, *That Man Haupt* (1973), is an excellent biography of a nineteenth century civil engineer, tunnel builder, and Civil War general. A second book by James Ward, *J. Edgar Thomson* (1980), reviews the work of a major president of the Pennsylvania Railroad. Patricia J. Davis, *End of the Line, Alexander J. Cassatt and the Pennsylvania Railroad* (1978), is an excellent life of the seventh president of the Pennsylvania. Brief essays on more than two dozen railroad leaders are discussed in Julius Grodinsky, *Transcontinental Railway Strategy, 1869–1893: A Study of Businessmen* (1962).

Suggested Reading

There is a wealth of material on early American transportation history. By far the best place to start is with Volume IV of *The Economic History of the United States*, which is George R. Taylor, *The Transportation Revolution, 1815–1860* (1951). Taylor gives a full and excellent account of internal improvements from roads through early rail development. Also useful is B. H. Meyer, Caroline E. MacGill, *et al.*, *History of Transportation in the United States before 1860* (1917). Transportation changes in the Carolinas and Georgia are reviewed in U. B. Phillips, *A History of Transportation in the Eastern Cotton Belt to 1860* (1913). For two good state studies, see Wheaton J. Lane, *From Indian Trail to Iron Horse: Travel and Transportation in New Jersey, 1629–1860* (1939), and John H. Krenkel, *Illinois Internal Improvements, 1818–1848* (1958). The substantial contribution made by the Army Engineers is reviewed in Forest G. Hill, *Roads, Rails and Waterways* (1957).

The early growth of New York is superbly shown in R. G. Albion, *The Rise of New York Port, 1815–1860* (1939). The trade competition of two other cities is seen in *The Philadelphia–Baltimore Trade Rivalry, 1780–1860* (1947) by James W. Livingood. A popular story of canals is Alvin F. Harlow, *Old Towpaths: The Story of the American Canal Era* (1926). *Steamboats on the Western Rivers* (1949) by Louis C. Hunter is another sound book on river transport. Several chapters in Robert E. Riegel, *Young America, 1830–1840* (1949), give delightful coverage of competing turnpikes, canals, steamboats, and railroads in the early nineteenth century.

Much of the material on early railroads is found in scholarly journals. Robert S. Cotterill has three articles on southern rail development in the *Mississippi Valley Historical Review:* "Southern Railroads and Western Trade, 1840–1850," III (March 1917), 427–41; "The Beginnings of Railroads in the Southwest," VIII (March 1922), 318–26; and "Southern Railroads, 1850–1860," X (March 1924), 396–405. See also Charles W. Turner, "Virginia

Railroad Development, 1845–1860," *The Historian* (Autumn 1947), 43–62. Western lines in the same years are reviewed in Robert E. Riegel, "Trans-Mississippi Railroads during the Fifties," *Mississippi Valley Historical Review*, X (September 1923), 153–72. Harry H. Pierce has a scholarly study of state aid for railroads in *Railroads of New York: A Study of Government Aid, 1826–1875* (1953). A small war between farmers and the railroads in Michigan is graphically presented in Charles Hirschfeld, *The Great Railroad Conspiracy* (1953). Mid-century urban rivalry is found in Wyatt W. Belcher, *The Economic Rivalry between St. Louis and Chicago, 1850–1880* (1947).

For a better understanding of the rail network on the eve of the Civil War, see the valuable maps and text in George R. Taylor and Irene D. Neu, *The American Railroad Network, 1861–1890* (1956). Northern railroads on the eve of war are well described in Carl R. Fish, "The Northern Railroads, April, 1861," *American Historical Review*, XXII (July 1917), 778–93. See also Herman K. Murphey, "The Northern Railroads and the Civil War," *Mississippi Valley Historical Review*, V (December 1918), 324–38. Longer and equally good is Thomas Weber, *The Northern Railroads in the Civil War, 1861–1865* (1952). The significant role of a border railroad is presented in *The Baltimore and Ohio in the Civil War* (1939) by Festus P. Summers. George E. Turner, *Victory Rode the Rails* (1953), reviews the strategic place of both Union and Confederate roads in the Civil War. By far the best work on Confederate railroads is Robert C. Black, *The Railroads of the Confederacy* (1952). Also useful are the two articles, C. W. Ramsdell, "The Confederate Government and the Railroads," *American Historical Review*, XXII (July 1917), 794–810; and Robert E. Riegel, "Federal Operation of Southern Railroads," *Mississippi Valley Historical Review*, IX (September 1922), 126–38.

Several recent books review railroad history before the Civil War. Albert Fishlow, *American Railroads and the Transformation of the Ante-Bellum-Economy* (1965), is an excellent survey of the effect of railroads on the American economy prior to 1860. *Boston Capitalists and Western Railroads* (1967), by Arthur M. Johnson and Barry E. Supple, is a detailed account of the leadership of Boston bankers as they projected early western railroads. James D. Dilts, *The Great Road* (1993), is a carefully crafted review of the early construction of the Baltimore & Ohio Railroad, 1828–1853. Allen W. Trelease, *The North Carolina Railroad, 1849–1871* (1991), is a definitive story of the first two decades of an important southern railway. Eugene Alvarez, *Travel on Southern Antebellum Railroads, 1828–1860* (1974), is a colorful review of early rail travel in the South. The marked increase in railroad mileage in the decade of the 1850s is outlined in John F. Stover, *Iron Roads to the West* (1978).

1865–1917

Western rail construction is well presented in Glen C. Quiett, *They Built the West: An Epic of Rails and Cities* (1934). Still of interest for a scandal in western rail expansion is Jay B. Crawford, *The Credit Mobilier of America* (1880). Written from the point of view of a railroad man is Robert S. Henry's extremely valuable "The Railroad Land Grant Legend in American History Texts," *Mississippi Valley Historical Review*, XXXII (September 1945), 171–94. Pertinent comments on the latter article by several qualified historians can be found in *Mississippi Valley Historical Review*, XXXII (March 1946), 557–76. See also Carter Goodrich, *Government Promotion of American Canals and Railroads, 1800–1890* (1960). The colorful history of several short mining railroads in Nevada and California is covered in Gilbert H. Kneiss, *Bonanza Railroads* (1941).

Though marred by its muckraking emphasis, Matthew Josephson, *The Robber Barons: The Great American Capitalists, 1861–1901* (1934), is very useful for the period of railroad corruption and the regulation which followed. Also valuable are chapters in two volumes of *A History of American Life:* Allan Nevins, *The Emergence of*

Modern America, 1865–1878 (1927), and Ida M. Tarbell, *The Nationalizing of Business, 1878–1898* (1936). For the railroads of the South just after the Civil War, see Carl R. Fish, *The Restoration of the Southern Railroads* (1919). John F. Stover, *The Railroads of the South, 1865–1900* (1955), traces the growing northern financial control over the southern lines. Northern rail problems and corruption are covered in C. F. Adams, Jr., *Railroads: Their Origin and Problems* (1880), and C. F. Adams, Jr., and Henry Adams, *Chapters of Erie and Other Essays* (1886). Lee Benson, *Merchants, Farmers and Railroads: Railroad Regulation and New York Politics, 1850–1887* (1955), is a scholarly work emphasizing the urban origins of rail regulation. A lively account of railroad labor violence can be found in Robert V. Bruce, *1877: Year of Violence* (1959).

An excellent introduction to the problems of distance and the high costs of transportation which faced the western farmer can be found in chapters of Frederick L. Paxson, *History of the American Frontier, 1763–1893* (1924). Also valuable for postwar farm problems is Fred A. Shannon, *The Farmer's Last Frontier: Agriculture, 1860–1897* (1945). Specific details of the western farmer's fight with the railroads are found in S. J. Buck, *The Granger Movement* (1913), and, by the same author, *The Agrarian Crusade* (1920).

Some of the first important aspects of railroad integration in the late nineteenth century are fully treated in a volume mentioned earlier, George R. Taylor and Irene D. Neu, *The American Railroad Network, 1861–1890* (1956). A brief pamphlet by Carlton J. Corliss, *The Day of Two Noons* (1952), reviews the introduction of standard time by the railroads. The effects of the Panic of 1893 on railways is covered in Edward G. Campbell, *The Reorganization of the American Railroad System, 1893–1900* (1938). An excellent discussion of the rail system in the early twentieth century and the challenges it faced from new competitive facilities is found in several chapters in Harold U. Faulkner, *The Decline of Laissez Faire, 1897–1917* (1951).

Suggested Reading

Delores Greenberg, *Financiers and Railroads, 1869–1889: A Study of Morton, Bliss & Company* (1980), is the story of a late nineteenth century banking firm and its railroad investments. An excellent account of the involved railroad relationships in the South in the 1880s is Maury Klein, *The Great Richmond Terminal: A Study of Businessmen and Business Strategy* (1970). H. Craig Miner, *The St. Louis-San Francisco Transcontinental Railroad: The Thirty-fifth Parallel Project, 1853–1890* (1972), reviews the early history of an important southwestern line. Three interesting volumes on railroad labor are: James H. Ducker, *Men of the Steel Rails* (1983); James A. Ward (ed.), *Southern Railroad Man* (1970); and Richard Reinhardt (ed.), *Workin' on the Railroad* (1970). K. Austin Kerr, *American Railroad Politics, 1914–1920* (1968), covers America's railroad problems during World War I, the federal control of the lines, and the Transportation Act of 1920. *Enterprise Denied* (1971), by Albro Martin, is an intriguing and first-class account of how American railroads were overregulated by the federal government between 1897 and 1917.

SINCE 1917

Relatively little has been written on the American railroad development of the twentieth century. For the rail crisis of 1917, see William G. McAdoo, *Crowded Years* (1931). Several chapters of his reminiscences are extremely useful for a study of the government railroad operation from 1917 to 1920. Walker D. Hines, *War History of American Railroads* (1928), while biased, is still essential for an understanding of the wartime operation. I. L. Sharfman, *The Interstate Commerce Commission* (4 vols., 1931–37), is the standard work on the subject. A shorter treatment is included in R. E. Cushman, *The Independent Regulatory Commissions* (1941). A major work on governmental aid is Federal Coordinator of Transportation, *Public Aids to Transportation* (4 vols., 1938–40). For railroads in the Second World War, see Joseph R. Rose, *American Wartime Trans-*

285

portation (1953), and also Carl R. Gray, Jr., *Railroading in Eighteen Countries* (1955). A sound comprehensive study of American rail problems since 1945 is James C. Nelson, *Railroad Transportation and Public Policy* (1959). Robert G. Lewis, *The Handbook of American Railroads* (1956), is a valuable source book of basic facts for the Class I railroads that are presently serving the nation.

Several books have covered railroad developments since the 1960s. Robert B. Carson, *Main Line to Oblivion* (1971), reviews in detail the decline of the railroad industry in the state of New York. The revitalization of an important southwestern road is clearly told in H. Craig Miner, *The Rebirth of the Missouri Pacific* (1983). H. Roger Grant, *Erie-Lackawanna: Death of an American Railroad, 1938–1992* (1994), describes the failure of the merger of two eastern lines. *Diesels West* (1963) by David P. Morgan is a richly illustrated story of diesel power on the Burlington Lines. Stephen Salsbury, *No Way To Run A Railroad, The Untold Story of the Penn Central Crisis* (1982), is a definitive history of the Penn Central merger and bankruptcy. Two detailed volumes on railroad mergers are Michael Conant, *Railroad Mergers and Abandonments* (1964), and Richard Saunders, *The Railroad Mergers and the Coming of Conrail* (1978). Claiborne Pell, *Megalopolis Unbound* (1966), looks at the future of railroad transportation for the supercity and the Northeast Corridor. *The Amtrak Story* (1994) by Frank N. Wilner is a fully documented history of the government's effort to provide rail passenger service for the nation.

Index

Index

Index

Index

Grant, Ulysses S., 69, 73, 89, 118
Gray, Carl R., 212, 215
Gray, Carl R., Jr., 190
"Great Eastern Mail," 32
Great Northern Ry.: construction of, 74; mentioned, 97, 126, 128, 199, 202, 212, 239
Great Plains: railroads help to populate, 61–62; different from earlier frontiers, 62–63; railroads in, 78–80
Great Western R.R., 153
Greeley, Horace, 61
Green Line (fast-freight), 140
Greenough, Allen J., 234
Greyhound System, 194
Guaranteed annual wage, 223
Guidebooks (railroad), 48
Gulf Mobile & Ohio R.R., 240, 250
Gurley, Fred H., 217
Guthrie, James, 42, 52

Hale, Nathan, 14
Hall, Harold H., 246
Hall, John M., 136
Hannibal & St. Joseph R.R., 58, 87, 161
Harnden, William F., 32
Harriman, Edward H., 107, 126, 128, 136, 165
Harriman lines (in 1906), 126
Harrisburg, Pa., 27, 221
Harrison, Benjamin, 142
Harrison, Fairfax, 271
Harrison, Joseph, 25
Hartford and New Haven R.R., 136
Harvey, Frederick H., and Harvey Houses, 75–76
Haupt, Herman, 18, 57, 59
Hayes, Rutherford B., 76
Headlight, invention of, 25
Heating of passenger cars, 154
Heineman, Ben W., 240
Hepburn, Alonzo B., and investigation of rail abuses, 118

Hepburn Act (1906), 115, 130, 168
Hepburn Committee, 111
"High-level" trains, 217
High-Speed Ground Transportation Act of 1965, 237
Highway freight business, 195–96, 218–19
Hill, James Jerome: first interest in railroads, 73; and Great Northern, 74; "Empire Builder," 74, and Northern Securities Co., 128–29; mentioned, 81, 97, 166, 239
Hill roads (in 1906), 126
Hines, Walker D., 174, 175, 180
Holden, W. W., 99
Holliday, Cyrus K., 74–75
Homestead Act, 83
Hopkins, Mark, 67, 68, 71, 76
Hopper cars, 152
Horses, as early motive power for railroads, 12–13, 24
Horton, Mrs. Silas, 103
Household goods, shipment of lost by railroad to truck, 195
Howe, Elias, 156
Howe, William, 156
"Howe truss" bridge, 45, 156
Hoxie, Herbert M., 65
Hudson and Manhattan R.R., 174
Hudson River: steamboat service, 8; mentioned, 15, 33, 101
Hudson's Bay Co., 73
Huntington, Collis P.: shares in organizing Central Pacific, 67; financial and political agent for Central Pacific, 68; and Southern Pacific, 76; death of, 128, 165; mentioned, 71

Ice, halts in navigation, 8, 9, 32
Illinois: early railroads in, 40–42; and Granger laws, 119; mentioned, 29, 36, 39, 45, 82, 88, 91, 99, 146
Illinois Central R.R.: early construction of, 41–42; receives land grant,

Index

Index

Maine Central R.R., 136
Mallet, Anatole, 151
Mallet articulated locomotive, 151
Mann-Elkins Act (1910), 130
Mansfield, Josie, 105
Manual of the Railroads of the United States, 37
Manufactured goods, shipment of lost by railroad to truck, 196
Marietta and Cincinnati R.R., 40
Markham, Charles H., 207
Marx, Karl, 105
Mason, William, 150
Massachusetts: first railroads, 14–15; mentioned, 19, 26, 155
Master Car-Builders Association, 142
Mather, Alonzo, 152
Mather Palace Stock Car, 152
Meigs, Return J., Jr., 4
Menk, John M., 239
Memphis and Charleston R.R., 43
Metroliners, 256
Michigan Central R.R.: completed to Chicago, 40; mentioned, 28, 80, 102, 153
Michigan Southern R.R., 40
"Midnight tariffs," 128
Mileage, railroad: in 1840, 19, 20; in 1850, 25–28, 35; in 1860, 35, 36, 43, 48–49; in 1865, 59; western, 77–78; and land grants, 83; and narrow gauge, 88–89; in prairie states, 91, 94; as compared to population, 94; in South, 97–98; post–Civil War, 133–35; as of 1916, 192; decline of, 192, 225; growth and decline of (1840–1959), 204–5; U.S. total (1830–1959), 205; (since 1960), 240, 257–58
Military Railway Service, 190
Milk shipments, 33
Milwaukee and Mississippi R.R., 81
Miner's frontier, and western railroads, 89–90, 95

Minor, D. Kimball, 20
Minot, Charles, 38
Mississippi and Ohio Ry., 157
Mississippi Central R.R., 42
Mississippi River: railroad bridge across, 41, 160; mentioned, 35, 159
Missouri-Kansas-Texas R.R., 88, 247
Missouri Pacific R.R., 87, 105, 126, 147, 151, 202, 218
Mitchell, Alexander, 81, 120
Mobile and Ohio R.R., 29, 42, 82, 143
Mogul-type locomotive, 44, 150
Mohawk and Hudson R.R., 19, 24, 26, 41
Monon R.R., 217
Monopoly of railroads, 164–65
Moore, William H., 127
Morgan, John Pierpont: and New York Central–Pennsylvania dispute, 110; controls many railroads by 1906, 126; and Southern Ry., 138; mentioned, 128, 136, 165
Morgan roads (in 1906), 126
Morris Canal, 6
Morrison, William R., 124
Morse, Wayne L., 189
Motive power: early improvements in, 24–25; improvements after Civil War, 147–51; mentioned, 157–58, 193
Mountain West: railroads help to populate, 61–62; nature of railroads in, 78; and narrow-gauge lines, 89–90, 143
Municipal aid for railroads, 29
Munn v. Illinois, 121

Name trains (passenger), 215–16
Narrow-gauge railroads, 88–90
Nashville, Chattanooga, and St. Louis R.R., 207, 231
"Natchez" (steamboat), 159
National Railroad Passenger Corporation, 234
National Road, 2, 4

Index

Index

Index

Index

Index